UNDERSTANDING BIOSCIENCES

UNDERSTANDING BIOSCIENCES

By

Dr. Ashok Kumar

Dept. of Zoology
Bundelkhand University
Campus Department
Jhansi

DISCOVERY PUBLISHING HOUSE PVT. LTD.
NEW DELHI-110 002

First Published-2010

ISBN 978-81-8356-529-5

Published by:

DISCOVERY PUBLISHING HOUSE PVT. LTD.

4831/24, Ansari Road, Prahlad Street
Darya Ganj, New Delhi-110002 (India)
Phone: 23279245 • Fax: 91-11-23253475
E-mail: parul.wasan@gmail.com
info@discoverypublishinggroup.com
Website: www.discoverypublishinggroup.com

Printed at:

Sachin Printers
Delhi

Preface

The present title "Understanding Biosciences" has been written for those students interested in careers in diverse fields of biological sciences. It provides a structured approach to learning by covering all the important topics in a uniform, systematic format. The book has been comprehensively designed incorporating recent advances in this fast moving field. It also provides accessible information on biosciences in compact form for undergraduate students in biology and related life sciences. It is intelligible to the educated layman, though it deals with some complex ideas. It is an adequate text for all the requirements of students in this area. In addition, busy lecturers who require a quick reference compendium will find it useful, particularly for tutional planning. Simple, yet hopefully clear figures and tables are provided throughout the book.

The over-riding goal of this book, and indeed of the whole *Understanding series*, is to present the essential information concering biosciences in a compact, readily accessible form which leads itself to student learning and revision. The convergence of various approaches has generated a rich panorama of detail, the significance of which we are still attempting to unraval. The present text has been written as an introduction to this rapidly growing field.

To make the work more comprehensive and informative, the author has consulted many authoritative books, research journals, abstracts, monographs etc., so there can be no claim to originality except in the manner of treatment.

The author expresses his thanks to his friends and colleagues whose continue inspirations have initiated him to bring out this book.

The author expresses his gratitude to Mr. Wasan and staff of M/s Discovery Publishing House Pvt. Ltd. for their whole hearted co-operation in the publication of this book.

In the mean time, the author will remain sincerely responsible for any shortcomings of the book and be grateful to the readers for their suggestions and constructive criticism for the continuous betterment of the book. He takes this opportunity to appeal to the readers to send their suggestions straightaway to his Publisher.

Author

Contents

1

INTRODUCTION

PRINCIPLES OF CLASSIFICATION

Underlying Principles

Early life forms were simple prokaryotic cells (bacteria and blue-green algae) which, through the slow process of adaptation to new environmental conditions, evolved into millions of organisms, each sufficiently different to prevent them breeding with each other.

It was inevitable that man should attempt to name these organisms and form them into an ordered *hierarchy*. Initially this was achieved on the basis of morphology and behaviour with the simplest at the bottom and man himself at the top (Aristotle's *Scala Naturae)*.

With the gradual acceptance that all species arose by adaptation of existing forms, the basis of this hierarchy became evolutionary, and simple and complex forms were placed in the same group because they shared a common *ancestry*.

Species are groups that have diverged most recently, genera some time earlier and so on up the taxonomic ranks. The difficult task of determining an organism's ancestry forms the basis of the science of classification.

Points of Perspective

The historical background to taxonomy including the contributions made by Aristotle (384322 BC), Ray (1628-1705), Linnaeus (1707-1778) and Darwin (1809-1882).

Essential Information

The science of biological classification is called *taxonomy*.

Reasons for Classification

There are over two million species which must be organized into groups before any useful study of them can be made. A good universal system of classification aids *communication* between scientists and allows information about a particular organism to be found more readily (e.g. in libraries).

Organisms form a *continuum* and any *grouping* of them is arbitrary and devised only for the convenience of man. Because evolution is a continuous process and new species are still being discovered, it is sometimes necessary to revise classifications.

What is a species?

Species are groups of interbreeding populations which are reproductively isolated from other such groups. The *biological* definition of a species emphasizes its genetic isolation. *Morphological* distinctiveness provides only a general guide for *delimiting* species.

Classification by Differences—Artificial Classification

This takes two forms:

1. Arranging organisms in groups according to whether or not they possess a particular characteristic and then subdividing on the basis of presence or absence of another characteristic.
2. Removing from the group of organisms all those with an obvious peculiarity to one set and those with another to a second set and so on until a large number of mutually exclusive groups is formed.

These two methods rely at each division on the use of a single character and will not only group together unrelated forms but a situation may be reached where no obvious diagnoses can be made at all.

However, the end-point of mutually exclusive groups makes these two methods useful for keys and the *dichotomous* key produced by the first system is most commonly used for identifying organisms.

Classification by Similarities—Natural Classification

The need for this was recognized as early as the eighteenth century by the taxonomist Linnaeus. Organisms are divided into groups and subgroups according to their basic similarities. Linnaeus worked before evolution was an accepted concept.

In order to develop a natural classification today taxonomists use *morphological*, *anatomical*, *biochemical*, *hybridization* and *chromosome* studies to arrive at tentative conclusions about phylogenetic (evolutionary) relationships between species. Such relationships must be based on homologous not analogous characteristics.

Homologous Characteristics

An organ of one organism is said to be homologous with an organ of another if both have a fundamental similarity of origin, structure and position which is

manifested especially during embryonic development, regardless of their functions in the adult, e.g. mammalian ear ossicles are homologous with the bone concerned with jaw attachment in a fish.

Analogous Characteristics

An organ of one species is said to be analogous to an organ of another when both organs have the same function and when they are not homologous, e.g. wings of birds and butterflies.

Rank

It is convenient to distinguish large groups of organisms from smaller ones; a series of rank-names is used which allows one to estimate the position of any particular group in the natural hierarchy. Linnaeus was the first to devise a comprehensive scheme and the rank-names in use today derive mainly from him.

Organisms are divided into large groups, phyla, each showing a radically different body plan; diversity within each allows the phylum to be divided into classes. Each class is divided into orders of organisms which have additional features in common.

Orders are subdivided into families and at this level differences are less obvious. Each family is divided into genera and each genus into species.

Binomial Nomenclature

Every organism is given a scientific name according to an internationally accepted system of nomenclature, first devised by Linnaeus. The name is always in Latin and is in two parts. The first name indicates the genus and is written with an initial capital letter; the second name indicates the species and is written with a small initial letter. These names are always distinguished in text by the use of italics or by underlining.

Rank	**Man**	**Sweet Pea**
Phylum	Chordata	Spermatophyta
Class	Mammalia	Angiospermae
Order	Primates	Rosales

Family	Hominidae	Leguminosae
Genus	*Homo*	*Lathyrus*
Species	*sapiens*	*odoratus*

Listed above are the obligate ranks of classification to which every specimen must be assigned but a *taxonomist* may use any number of additional *categories* within this scheme. The principle rank-names in use today are listed below in *hierarchical*.order.

Kingdom, Subkingdom, Grade, *Phylum*, Subphylum, Superclass, *Class*,. Subclass, Infraclass, Superorder, *Order*, Suborder, Infraorder, Superfamily, **Family**, Subfamily, Tribe, *Genus*, Subgenus, *Species*, Subspecies, Variety.

VIRUSES, MONEA, POTISTA AND FUNGI

Underlying Principles

As living organisms evolved from non-living material, it is not surprising that the dividing line between some of them is thin. Viruses, for example, possess features of both living and nonliving material. The other acellular organisms are far from being the simple structures they are sometimes thought to be.

Within the confines of a single cell they show remarkable specialization of structures to suit a widely divergent range of environments. They are the oldest organisms in the world (3000 million years old); indeed more than three-quarters of the time since life first evolved on earth was occupied by acellular organisms alone.

They are also the most abundant; one gram of fertile soil may contain as many as 2500 million bacteria.

Points of Perspective

Differences between prokaryotic and eukaryotic cells and the theory of evolution of eukaryotic cell organelles from prokaryotic cells. Importance of early organisms in changing the composition of the earth's atmosphere, in particular in the production of free oxygen.

Historical aspects of the control of disease caused by pathogenic protista, in particular the work of Jenner (1748-1823), Pasteur (1822-1895), Koch (1843- 1910), Lister (1827-1912) and Fleming (1881-1955).

Essential Information

In the past all organisms have been classified as plants or animals according to whether they possess chlorophyll and photosynthesize, as do plants, or lack chlorophyll and obtain their food from other organisms, as do animals. Many *acellular* organisms are capable of both modes of life. It is, therefore, convenient to distinguish other kingdoms of organisms, distinct from both plants and animals.

For the purposes of this book *Viruses*, *Monera*, *Protista* and *Fungi* will be considered as separate kingdoms.

Viruses exist outside living cells where they may be considered non-living chemicals. They enter living cells and once inside, multiply with the assistance of the host cells. Viruses contain very few enzymes and so, as *intracellular parasites*, they use the host's enzymes for their own metabolism.

The presence of viruses was first postulated by the Russian Iwanowski at the end of the nineteenth century and they were isolated in crystalline form in 1935 by the American biochemist Stanley.

Viruses range in size from 20 nm-300 nm, i.e. below the resolving power of the light microscope. They are highly specific to both host and a particular tissue within a host.

They are generally classified according to the host they infect, i.e. plant viruses, animal viruses and bacterial viruses (bacteriophages). Essentially viruses are composed of a core of nucleic acid surrounded by a protein sheath.

In some, e.g. fowl plague virus, fat and carbohydrate are also present. The nucleic acid in bacteriophages is always DNA; plant viruses all contain RNA and animal viruses may contain DNA or RNA.

There are many shapes of virus but they are frequently rod-shaped or polyhedral.

Life cycle of a *virulent phage*, e.g. T-phage of *Escherichia coli*

1. The virus becomes adsorbed onto the surface of *E. coli* by fibrils.
2. The needle is pushed into the *bacterium* and the DNA moves into the *host cell*.
3. Immediately replication of bacterial DNA (and therefore enzyme production) ceases.
4. All systems of the host cell are taken over by viral DNA and used to produce more strands of viral DNA.
5. The host enzyme and synthetic systems are used to produce protein coats etc. for this DNA. More than 100 new viruses are produced in this way.
6. Viral DNA causes the host systems to produce enzymes which

cause lysis of the host cell wall, thus releasing the new viruses. The whole process generally takes about one hour, but may vary depending on the virus and environmental conditions.

Life cycle of a temperate phage

The process is much less rapid and the host and phage may exist together for many generations. Host DNA may become incorporated in the viral DNA, and this bacterial DNA is carried to the next host, thereby resulting in new characteristics. This process of *transduction* is an important method by which antibiotic-resistance spreads throughout a population of bacteria.

Diseases caused by viruses

Plant viruses cause mosaic diseases, e.g. of tobacco, potato, bean and sugar cane. Animal viruses cause *fowl plague*, *mumps*, *cold sores*, *influenza*, *smallpox* and *poliomyelitis*.

Monera

Their group includes bacteria and blue-green algae.

Bacteria

Structure

Bacteria range in size from 0.5μm-4.0μm.

Their shape may be spherical (coccus), rod-like (bacillus) or spiral (spirillum). The spherical forms may occur in chains (streptococcus) or in clusters (staphylococcus).

They are the smallest form of life organized on a cellular basis but the cells differ in a number of respects from those of other organisms. They are known as prokaryotic cells and have the following characteristics:

1. no distinct nucleus
2. DNA not incorporated in chromosomes but comprising a single, circular strand
3. no spindle forms at cell division
4. membrane-bounded organelles (e.g. Golgi, endoplasmic reticulum, mitochondria) are absent
5. no large vacuoles
6. cell wall of protein and polysaccharide (composition varies with genera).

Mode of life

Bacteria are often classified on the basis of their method of

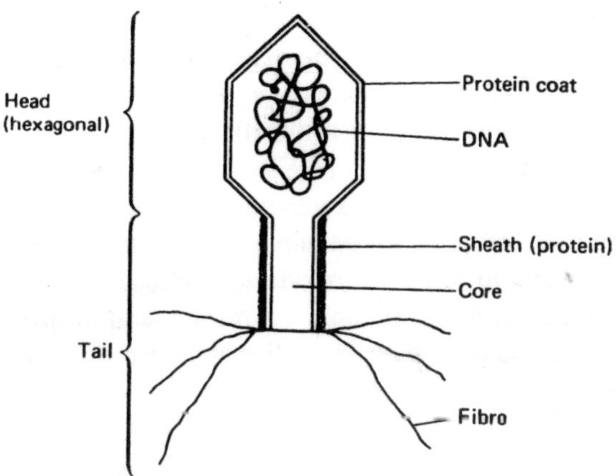

Figure 1.1: (a) Structure of Bacteriophage.

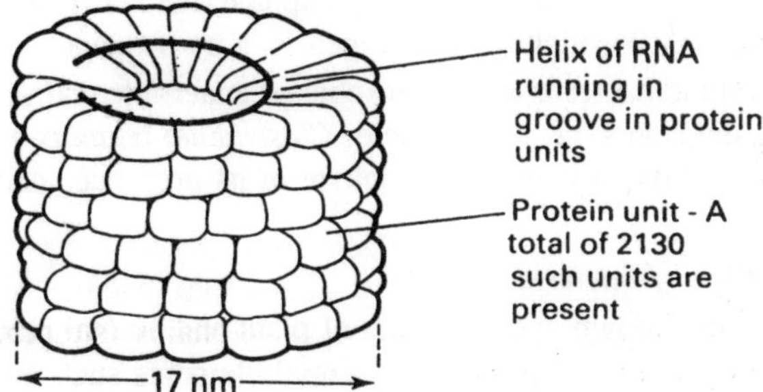

Figure 1.1: (b)

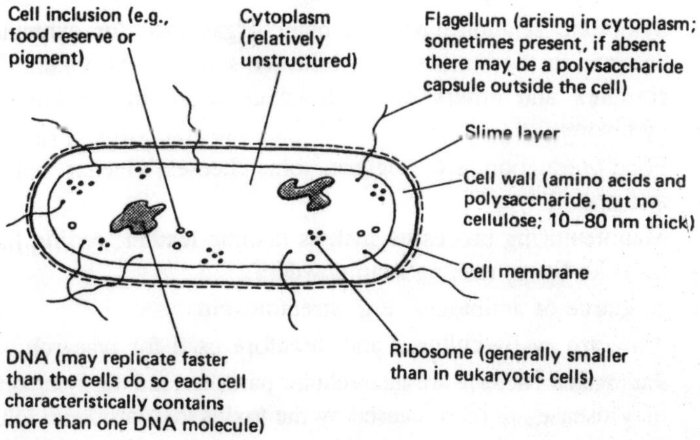

Figure 1.1: (c) Generalized bacterial cell.

obtaining energy:

1. Most obtain energy from the oxidation or breakdown of living or non-living organic matter, i.e. are heterotrophic. They are parasitic, saprophytic or symbiotic.
2. A few obtain energy from the oxidation of inorganic materials and use this energy to synthesize their own foods, i.e. are chemo-autotrophic; for example, iron bacteria oxidize ferrous compounds to ferric hydroxide and release energy; nitrifying bacteria oxidize ammonia to nitrate and release energy; colourless sulphur bacteria oxidize hydrogen sulphide to sulphur and release energy.
3. Some bacteria can use the energy of sunlight for the manufacture of food, i.e. are photoautotrophic; for example, green and purple sulphur bacteria contain bacteriochlorophyll and photosynthesize using hydrogen sulphide (not water) as a source of hydrogen; sulphur, not oxygen, is a by-product.

Heterotrophic bacteria may be aerobic or anaerobic and some may live in either environment, for example, *Clostridium tetani is* a parasite in the absence of oxygen and a saprophyte in its presence. Autotrophs are anaerobic.

The importance of bacteria

1. The breakdown and recycling of plant and animal remains, in particular the recycling of essential elements such as carbon, nitrogen and phosphorus. The same processes account for the bacterial decomposition of sewage.
2. Symbiotic relationships with other organisms; for example, bacteria in the human gut synthesize some of the vitamin B complex and others break down cellulose in the guts of herbivores.
3. Food production, e.g. yoghurt, some cheeses, vinegar, coffee and tea.
4. Manufacturing processes such as tanning leather, retting flax to make linen, making soap powders.
5. A source of antibiotics, e.g. streptomycin.
6. They are easily cultured and therefore used for research.
7. Pathogenic bacteria are intercellular parasites and the symptoms of a disease are often caused by the toxins they produce. Only five genera of bacteria are thought to infect plants but they infect a wide range; for example *Xanthomonas phaseolus* causes

common blight of beans. Human diseases caused by bacteria include whooping cough *(Bordetella pertussis)*, some forms of pneumonia, leprosy, syphilis, tuberculosis, diphtheria, typhoid *(Salmonella typhi)*, cholera and scarlet fever.

Protista

These consist of single-celled eukaryotic organisms. There are two main groups: the Protozoa and the Euglenophyta.

Protozoa

Table 1.1: Classes of Protozoa.

Phylum		**Characteristics**
Protozoa (first animals)		Acellular (unicellular) Mostly microscopic (2μm-10μm long) Specialized organelles but no tissues or organs
Class	**Characteristics**	**Examples**
Rhizopoda (Sarcodina)	Move and feed by means of pseudopodia	*Amoeba*
Flagellata (Mastigophora)	Move using flagella; asexual reproduction by longitudinal binary fission	*Trypanosoma; Euglena*
Ciliata	Move using cilia; meganucleus and micronucleus; asexual reproduction by transverse binary fission	*Paramecium*

Estimates of the number of species vary from 1500-50 000. There are free-living, symbiotic and parasitic species which are widely distributed but only active in *aqueous media*. They are highly specialized with division of labour and specialization of *organelles* which has resulted in a wide variation of form and physiology.

The flagellates include *holophytic* and *holozoic* forms. Protozoa form a significant part of plankton. Two orders of rhizopods, the Radiolaria and the *Foraminifera*, have hard tests (shells) and over millions of years these have fallen to the bottom of the oceans to form deposits thousands of feet thick.

Changes in sea level have exposed some of these as *chalk* and limestone. Most animals have one or more protozoan parasites, some of which cause disease in man including: *Trypanosoma* causing sleeping sickness and *Plasmodium* causing malaria.

Fungi

About 80 000 species of fungi have been recognized, including 30 000 Ascomycetes and 25 000 Basidiomycetes. They help maintain soil fertility by recycling many important minerals and decomposing the organic matter of both plants and animals.

Industrial uses of fungi include the extraction of enzymes (e.g. invertases) and drugs (e.g. steroids and ergotamine); cheese making; baking and brewing. Fungi cause diseases in plants, e.g.

Puccinia graminis — rust of wheat

Phytophthora infestans — late blight of potatoes

A few fungi cause diseases in man, e.g.

Microsporum audouini — ringworm

Trichophyta interdigitale — athlete's foot

Stored food can be damaged by moulds; dry rots attack wooden construction materials and mildews affect cotton, wool and manufactured goods.

Table 1.2: Groups of Fungi.

Kingdom		**Characteristics**
		Fungi No chlorophyll; normally have hyphae, often grouped as a mycelium; hyphae are coenocytic, although they may have septa; cell wall mostly hemicellulose, chitin, lipid and protein
Division	**Characteristics**	**Examples**
Phycomycetes	Aseptate mycelium; asexual spore is a zoospore or sporangiospore; sexual	*Phytophthora* — causes late blight of potatoes; *Mucor* - large genus, mainly sapr-ophytic stage is a zygos-pore or oospore
Ascomycetes	Septate hyphae; asexual reproduction by conidia or budding; sexual	*Saccharomyces* (yeast) - simple, consisting of a single cell; *Penicillium*

	reproduction by ascospores	
Basidiomycetes	Septate hyphae; hyphae often massed into extensive three-dimensional structures (puffballs; toadstools; bracket fungi); sexual reproduction by basidiospores	*Agaricus* - mushroom

PLANTS

Underlying Principles

Attempts to classify plants have been made since Theophrastus (c. 300 BC), but prior to Linnaeus (1707-1778) there was little, if any, uniformity in the nomenclature of plants; it was not until after Darwin (1809-1882) that attempts were made to classify plants according to their evolutionary relationships.

Plants are grouped together according to their similarities of morphology, anatomy, embryology, biochemistry and reproductive behaviour. As the groups vary slightly depending on those aspects which the taxonomist considers most important, there are a number of different schemes of classification.

Points of Perspective

A background knowledge of other plant classification schemes is useful. So is a knowledge of the geological periods during which the groups arose and were dominant, and a knowledge of some fossil representatives of the groups.

Algae

Table 1.3: Characteristics of Algae.

Phylum	Characteristics
Algae	Plant body is a thallus, i.e. not differentiated into stem, root and leaves
	All or most cells contain chlorophyll Food-conducting tissue is rare

This is an artificial but convenient group which includes the aquatic, autotrophic, non-vascular plants. Sub-divisions are based mainly on structural and biochemical differences associated with photosynthesis.

There are over 7000 species of Chlorophyccae and 1100 species of Phaeophyceae. 'hhe algae are found almost everywhere except in sandy deserts and permanent ice and snow. As primary producers they are a major component of phytoplankton.

Derivatives of alginic acid, which is extracted from Phaeophyceae such as *Luminaria* and *Ascophyllum,* are used in the following commercial processes: as thickeners in food; in cosmetics; in latex formation; as gelling agents in confectionery, meat jellies and dental impression powders; as emulsifiers in ice cream, polishes, processed cheese and synthetic cream; as surface films on glazed tiles; as medical gauzes; and in some types of sausage casings.

Table 1.4: Classes of Algae.

Class	Characteristics	Examples
Chlorophyceae (green algae)	Chlorophyll is the main photosynthetic pigment; chiefly found in fresh water	*Spirogyra*-filamentous, chain of identical cells containing spiral chloroplasts; Ova (sea lettuce)flat thallus
Phaeophyceae (brown algae)	Multicellular; contain chlorophyll + fucoxanthin (a brown pigment); store the carbohydrate laminarin	*Ectocarpus* - filamentous, prostrate anchoring filaments and upright photosynthetic reproductive filaments; *Fucus* (common intertidal wrack)- holdfast, stipe and flattened frond

Carrageenin is used in much the same way as *alginic acid* and is extracted from red seaweeds (*Rhodophyceae*) such as *Gigartina* and *Chondrus crispus*. These algae are also the richest source of agar, which is used as a medium for algal, bacterial and fungal cultures.

The siliceous walls of diatoms form large deposits of diatomite which are mined and used as an *insulation* material and as a filtration aid (e.g. sugar refining; brewing). In coastal areas seaweeds may be used as a fertilizer and for animal fodder. Algae aid the *oxidation* of sewage and thus promote the growth of *aerobic bacteria*.

They are used in percolation beds in water purification plants but in reservoirs they cause problems by blocking filters and giving the water an unpleasant taste. They prevent erosion by binding the surface of the soil. They may be used as a measure of water pollution.

Bryophytes

Table 1.5: Classification of Bryophytes.

Phylum		**Characteristics**
Bryophyta		No filamentous forms (except one stage in mosses) Clear alternation of generations, with gametophyte independent, sporophyte dependent Gamete-producing cells surrounded by jacket of sterile vegetative cells No asexual spores produced
Class	**Characteristics**	**Examples**
Hepaticae (liverworts)	Thalloid gametophyte is flat ribbon or leafy shoot; found in moist, shady places	*Riccia; Marchantia; Pellia*
Musci (mosses)	More conspicuous; better able to withstand drought; gametophyte has two growth stages: filamentous protonema and upright plant with spirally arranged leaves; more complex capsule	*Polytrichum; Funaria; Mnium*

There are over 23 000 species of bryophytes of which about 9000 are liverworts and 14 500 are mosses. They are mostly terrestrial, but water is essential for fertilization. They occur on soil and the trunks of trees; many are fully aquatic.

Pteridophytes

Table 1.6: Classification of Pteridophytes.

Phylum	**Characteristics**
Pteridophyta	Clear alternation of generations Sporophyte is most prominent phase Gametophyte reduced to

		simple prothallus Sporophyte has roots, stems, leaves and simple vascular tissues

Class	**Characteristics**	**Examples**
Lycopodiales (club mosses)	Densely packed small leaves borne on branched stems	*Selaginella*-heterosporous *Lycopodium*-homosporous
Equisitales (horse-tails)	Whorls of small leaves on upright stems; strobili (cones) at apex	*Equisetum*
Filicales (true ferns)	Prominent frond-like leaves; often sporangia on lower surface; underground rhizomes	*Dryopteris; Polypodium*

There are 11 000 species of Filicales and 100 species of Lycopodiales; the class Equisitales has one genus with two dozen living species. Most Pteridophyta today are relatively small but of the numerous fossil forms many were forest trees.

Table 1.7: Classification of Spermatophytes.

Phylum	**Characteristics**
Spermatophyta (Spermaphyta) (seed plants)	Heterosporous: male spore is pollen grain, female spore is embryo sac Embryo sac enclosed in ovule Seeds formed after fertilization Vascular tissues complex Gametophyte very reduced

Class	**Characteristics**	**Examples**
Gymnospermae (naked seeds)	No protective case (e.g. ovary wall) around ovule; most have cones (not *Ginkgo* or	*Pinus-pine;* *Picea* - spruce; *Larix*-larch

	Taxus); no vessels, only tracheids in xylem; many have needle like leaves which are often evergreen	

Of the 6500 species of Gymnosperms, most are xerophytic. They are common in temperate regions where the soil water may be frozen. Conifers grow quickly to produce the soft wood often used in building and for inexpensive furniture.

However, their greatest importance is in pulping and paper manufacture; about 40% of the world's timber harvest is used for paper. Tannins obtained from the bark of some conifers (e.g. spruce) are used to prevent the growth of bacteria on leather, to prevent scaling in boilers and for clarifying wines and beers. By-products of the timber trade are resins, which are used in paints, varnishes and printing inks.

Table 1.8: Classification of Angiosperms.

Phylum		**Characteristics**
Angiospermae		Flowers produced; ovules develop inside ovary; ovary wall develops into fruit; vessels in xylem
Sub group	**Characteristics**	**Examples**
Monocotyledons	Embryo has a single cotyledon; leaves usually show parallel venation; flower parts typically in multiples of 3; stem contains scattered vascular bundles; no secondary growth	*Endymion non-scriptus* (bluebell) - family Liliaceae; *Triticum* (wheat) - family Graminae
Dicotyledons	Embryo has two cotyledons; leaves have a network of veins; flowers often in multiples of 4 or 5; stem contains ring of vascular bundles; secondary growth occurs	*Ulex* (gorse) - family Leguminosae; *Helianthus* (sunflower) - family Compositae: *Prunus domestica* (plum) - family Rosaceae

These are the dominant plants of the world today with over 250 000 species, of which 60 000 are monocotyledons and 190 000 are

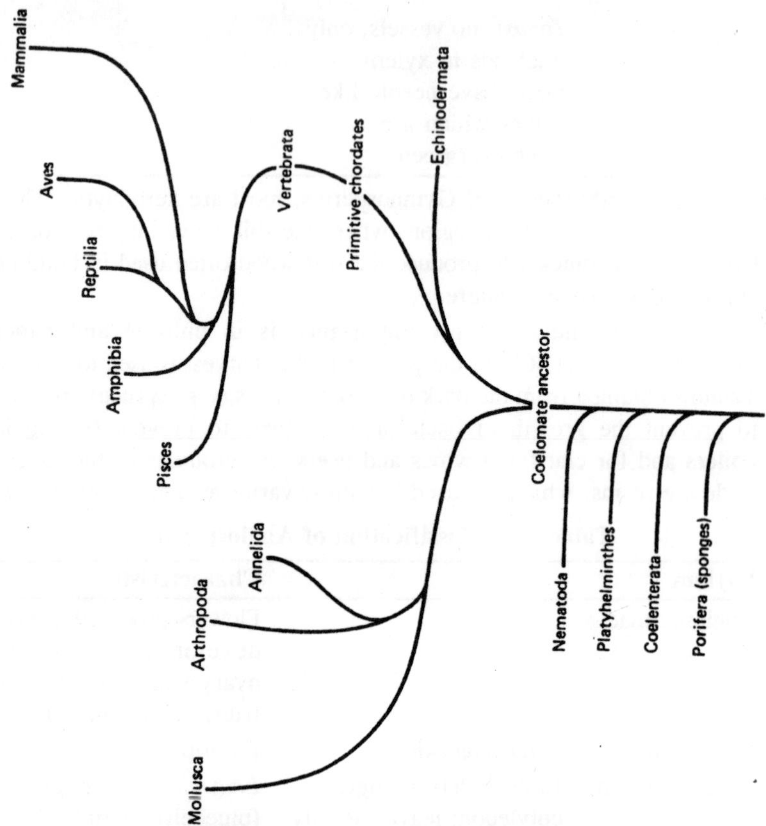

Figure 1.2: Evolutionary relationships between the major animal groups.

dicotyledons. They vary in form from simple duckweed *(Lemna)* through herbaceous and shrubby plants to trees such as chestnut and oak. They are the most important terrestrial primary producers.

Angiosperms are man's most important food plants: they include the cereals, vegetables, succulent fruits and sugar cane. As well as providing food for man's domesticated animals (pasture grasses and clover), they are a source of oils, drugs and insecticides.

The Leguminosae are very important for the nitrogen-fixing bacteria present in their root nodules. The angiosperms provide a variety of hardwoods.

Dicotyledons	Monocotyledons
Two cotyledons	One cotyledon
Net (reticulate) veined ieaves	Parallel veined leaves.

	Floral parts in groups of four or five (or Floral parts in threes (or multiples thereof) multiples thereof)
Vascular bundles in stem arranged cylindrically near periphery	Vascular bundles in stem scattered throughout-some central
Vascular cambium in all types showing	Little, if any, vascular cambium secondary growth

ANIMALS

Underlying Principles

If one accepts the Darwinian theory of evolution that each group has evolved from a preexisting one, the problem is to determine the order in which they arose and which groups are most closely related (phylogeny). The original ancestors of each group are now extinct, as are many of the intermediate groups.

Many soft-bodied forms have left no fossil record and others that have remain to be discovered. The picture of animal evolution is hence fragmentary and open to a number of interpretations. Figure elsewhere in this chapter shows one 'evolutionary tree' incorporating a number of possible theories. These theories have been based on the embryological and morphological relationships of the groups.

Points of Perspective

Theories of the origin of the Metazoa. The geological periods during which the groups arose and were dominant and a knowledge of a fossil example of each group.

Coelenterata

Table 1.9: Classification of Coelenterates.

Phylum	Characteristics
Coelenterata	Diploblastic, i.e. all cells derived from ectoderm or (hollow-intestine) endoderm Radially symmetrical Single body cavity (enteron; gastrovascular cavity) with only one opening to exterior Nematoblasts (stinging cells)

Class	Characteristics	Examples
		Polymorphism common (hydroid or medusoid forms)
Class	**Characteristics**	**Examples**
Hydrozoa	Typically hydroid and medusoid forms	*Hydra; Obelia; Physalia* - Portuguese man-of-war
Anthozoa	No medusoid stage	*Actinia* - sea-anemone; corals
Scyphozoa	Reduced hydroid stage	*Aurelia*-jellyfish

There are over 9000 species of coelenterates which are entirely aquatic and mostly marine. They feed only on living prey. Coral reefs form and protect certain islands and are useful for determining the age of rock deposits.

Platyhelminthes

Table 1.10: Classification of Platyhelminthes.

Phylum		Characteristics
Platyhelminthes (flatworms)		Triploblastic, i.e. cells derived from ectoderm, endoderm and mesoderm Bilaterally symmetrical (prerequisite for efficient locomotion) Dorso-ventrally flattened Acoelomate Gut, where present, branched with one opening
Class	**Characteristics**	**Examples**
Turbellaria	Free-living; ciliated epidermis	*Planaria*
Trematoda	Parasitic with one vertebrate and one invertebrate host	*Fasciola* - liver fluke; *Schistosoma*-blood fluke
Cestoda	Parasitic with two vertebrate hosts; body divided into proglottids	*Taenia* - tapeworm

The platyhelminthes include aquatic, free-living and parasitic forms. Many of them cause disease in man and/or in domesticated animals;

for example, bilharzia in man and liver rot in sheep are caused by *Schistosoma* and *Fasciola,* respectively.

Nematoda

Table 1.11: Classification of Nematodes.

Phylum	Characteristics
Nematoda (thread worms)	Unsegmented Cylindrical, worm-like shape Body cavity an unlined pseudocoel Mouth and anus present No circular muscles (movement restricted to dorso-ventral plane)

There are over 10 000 species of free-living and parasitic nematodes found in every conceivable niche from the poles to the tropics. Many cause disease in man, animals and crops, e.g. hookworm in *man-Ascaris;* pin-worm in *man-Enterobius; elephantiasis-Wuchereria (Filaria) bancroftii; Heterodera* is a parasite in the roots of tomatoes, cucumbers and beet.

Annelida

Table 1.12: Classification of Annelids.

Phylum		Characteristics
Annelida (ringed worms)		Coelom well developed (see below) Metamerically segmented (see below) Chaetae usually present Thin cuticle of collagen
Class	**Characteristics**	**Examples**
Oligochaeta	Obvious clitellum; few chaetae	*Lumbricus*-earthworm; *Allolobophora*-earthworm
Polychaeta	Many chaetae on parapodia	*Nereis* - ragworm; *Arenicola* - lugworm
Hirudinea	Ectoparasitic; suckers	*Hirudo*-leech

There are about 9000 species of freshwater, marine and terrestrial annelids which are mainly free-living. Earthworms improve soil fertility by increasing aeration, drainage and availability of nutrients. Leeches cause wounds through which pathogens may enter the body.

Importance of the Coelom

In coelomates the mesoderm is split into two layers divided by a fluid-filled cavity, the coelom. The development of the coelom allows the gut muscles to be separated from those of the body wall so that:

1. locomotion and digestion can take place independently and at the same time; this leads to the development of specialized locomotory and digestive organs.
2. a transport system (usually a blood system) must develop because the body cannot be flattened to produce a large surface area/volume ratio.
3. the presence of a transport system permits the development of specialized respiratory surfaces.

Metameric Segmentation

This is present in most coelomates. Basically the body of a metamerically segmented animal is made up of a series of identical segments of the same age. Each segment has a similar pattern of excretory organs, blood vessels etc., but they act independently (unlike a proglottis).

Often metamerism is difficult to detect in the adult, especially at the anterior end (due to cephalization) and near the limbs. The repetition of structures inherent in metameriszm lends itself readily to the adaptation of various parts of the body.

Arthropoda

Table 1.13: Classification of Arthropods.

Phylum		Characteristics
Arthropoda (jointed limbs)		Metamerically segmented Coelomate (haemocoel) Chitinous exoskeleton Jointed appendages Compound eyes often present (unique to arthropods)
Class	**Characteristics**	**Examples**
Crustacea	Exoskeleton hardened by calcite; Cephalothorax; 2 pairs antennae; Gills	*Daphnia* - water *flea* (subclass Branchiopoda) freshwater, parthenogenetic, filter feeder; *Oniscus*-woodlouse (subclass Malacostraca, order Isop-

		oda) terrestrial; *Astacus*-crayfish *Carcinus*-crab *Crangon* - shrimp
Chilopoda	Flattened body; 15-20 pairs long legs	*Lithobius* - centipede
Diplopoda	Cylindrical body; Up to 200 pairs short legs	*Julus* - millipede
Arachnida	Prosoma and opisthosoma (2 main body regions); 1 pair chelicerae; 1 pair pedipalps; 4 pairs walking legs; Lung books for respiration	*Araneus* - spider
Insecta	Head, thorax and abdomen; Exoskeleton strengthened by chitin and protein; 1 pair antennae; 3 pairs walking legs; Respiration by tracheal system	*Locusta* - locust (order Orthoptera) usually large herbivores, hind legs modified for jumping, stridulatory (sound producing) organs in male; *Libellula*-dragonfly (order Odonata) strong fliers, adult and nymph predatory carnivores; *Pieris brassica* - cabbage white butterfly (order Lepidoptera), scales on large wings, adults feed on liquids, larva is a caterpillar;*Musca domestic-a*-house-fly (order Diptera), hind wings modified to form halteres to control equili-brium in flight, legless larvae (maggots); *Apis mellifica* -honey-bee (order Hymenoptera) wide variety of mouthparts, many social forms

The Arthropoda is the largest and most successful animal phylum; over 800 000 species have been described. There are over 30 000 species of *Crustacea*, which are typically marine and form an important component of *zooplankton*; some are eaten by man.

The 3000 or so species of Chilopods are all fast-moving carnivores and the 9000 species of Diplopods are slow-moving herbivorous scavengers. The *Chilopoda* and *Diplopoda* were formerly placed in the Class Myriapoda. There are over 35 000 species of arachnids which are parasitic or free-living carnivores.

The early forms were mostly aquatic but present forms are mainly terrestrial. Many ticks carry disease. By far the most significant class of arthropods is the Insecta with more than 750 000 species described. Insects are mainly terrestrial and, although there are some freshwater forms, none are truly *marine*.

The insect body plan can become specialized for many different modes of life and therefore the group shows great adaptive radiation and is widely distributed because of the ability to fly. Insects can be beneficial or harmful to man.

Table 1.14: Beneficial and Harmful Insects.

Beneficial	Harmful
1. Use of insect products: silk; honey; bees' wax; cochineal; shellac	1. Destroy leaves and fruit of plants, e.g. locusts, boll weevils, fruit flies.
2. Commercial production of fruits dependent on insect pollination	2. Transmit disease, e.g. mosquitoes (malaria and yellow fever); house-flies (typhoid and dysentery); fleas (bubonic plague and typhoid); tsetse flies (sleeping sickness)
3. Biological control of harmful organisms, e.g. ladybirds eat aphids	3. Annoy or harm domestic animals, e.g. lice
4. Recycling of dead and decaying plant and animal remains, e.g. by beetles and flies	4. Cause destruction of materials in and around houses, e.g. cockroaches (spoil food); moth larvae (feed on carpets and clothing); termites (bite into wooden furnishings, or gnaw through the wooden supports of houses)
5. Food source of many economically important animals, e.g. trout	

Mollusca

Table 1.15: Classification of Molluscs.

Phylum		**Characteristics**
Mollusca (soft bodied)		Unsegmented Head, muscular foot and visceral mass Mantle secretes shell
Class	**Characteristics**	**Examples**
Gastropoda	Asymmetrical due to torsion; single shell, often coiled; feed using radula	*Littorina* - periwinkle; *Helix* - snail
Bivalvia	Shell in two halves; filter feeders	*Mytilus* - mussel; *Cardium* - cockle

There are over 80 000 living species of molluscs, which are typically marine but which have invaded freshwater habitats and land. Molluscs are valued as food for man, for the purple dye that can be obtained from *Murex,* and for pearls from oysters.

Deleterious effects include damage done by the shipworm *Teredo* to wooden jetties; furthermore slugs and snails are harmful to crops, and snails are intermediate hosts in the life cycles of some parasites of economic importance, e.g. *Schistosoma.*

Echinodermata

Table 1.16: Characteristics of Echinoderms.

Phylum	**Characteristics**
Echinodermata (spiny skinned)	Acquired radial symmetry (often pentamerous) No head Endoskeleton of calcareous ossicles with movable spines Watervascular system with tube-feet and madreporite

Examples

Asterias (Class Asteroidea) flattened, star-shaped flexible body;

Echinus - sea urchin

There are over 5000 living species of echinoderms, which are exclusively marine. The '*Star of Thorns*' starfish destroys coral.

2

BIOMOLECULES

CARBOHYDRATES, LIPIDS AND PROTEINS

Underlying Principles

Hydrolysis: the splitting of a molecule by the addition of water.

Condensation: the combination of two simple molecules to form a complex one with the release of water.

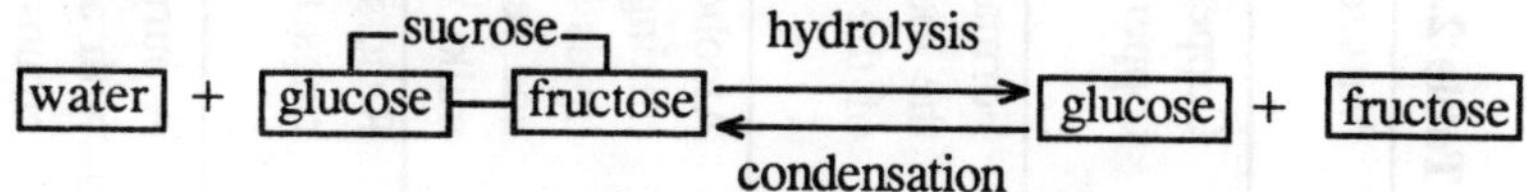

Polymers: long chains of similar units joined together to form a complex molecule.

Isomers: chemical compounds having the same number and kind of atoms, but differing in the arrangement of these atoms and hence in their properties.

Points of Perspective

The structure and function of carbohydrate derivatives and related substances, e.g. lignin, suberin, chitin and pectin, could be useful. It may help to have some knowledge of the nature of the changes that occur in each food test:

Carbohydrates

The general formula for carbohydrates is $(CH_2O)_n$

where n = 3 it is called a triose sugar (glyceraldehyde is $C_3H_6O_3$)

where n = 5 it is called a pentose sugar (ribose is $C_5H_{10}O_5$)

where n = 6 it is called a hexose sugar (glucose is $C_6H_{12}O_6$)

Table 2.1: Food Tests.

Test for	*Reagent(s)*	*Active constituent of reagent*	*Nature of the colour change*	*Additional notes*
Reducing and non-reducing sugars	Benedicts	Copper (II) sulphate	Copper (II) sulphate (blue)→ copper (1) oxide (orange/red)	Reaction only occurs in alkaline conditions
Starch	Iodine dissolved in potassium iodide solution	Iodine	Formation of a complex with the iodine trapped in the centre of the helical starch molecule	Heating above 70°C causes unwinding of the starch helix and loss of colour. Test must be carried out at room temperature or below
Lipids	Alcohol water	Alcohol	Alcohol dissolves the lipid. On mixing this with water the lipids become finely dispersed and refract light, giving a milky appearance.	Thorough mixing of water and lipid laden alcohol gives the best result. Solution goes clear in time due to settling out of the lipids.
Protein	Millons	Mercuric nitrate (in nitric acid)	Reacts with certain phenol rings to give a red coloration	Of the amino acids present in proteins, only tyrosine gives a positive result
	Biuret	Copper (II) sulphate	A purple complex is formed with adjacent pairs of $-CONH_2$ groups	If excess copper sulphate is added a blue precipitate of copper hydroxide is formed

These are all examples of sugars (saccharides). Where there is only one sugar it is called a monosaccharide; two monosaccharides can join together to form a disaccharide, and many monosaccharide units form polysaccharides.

Monosaccharides 'single sugars' e.g. glucose; fructose; galactose

All three have the same basic formula, $C_6H_{12}O_6$, but differ in the arrangement of the atoms, i.e. are isomers; for example, glucose is an aldose sugar whereas fructose is a ketose sugar. They all reduce Benedict's (Fehlings) Reagent, are sweet, soluble and easily transported, and are the main respiratory substrates.

Disaccharides 'double sugars': These are formed by the condensation of any two monosaccharides e.g.:

glucose + glucose = maltose (malt sugar)

glucose + fructose = sucrose (cane sugar)

glucose + galactose = lactose (milk sugar)

These three disaccharides have the basic formula $C_{12}H_{22}O_{11}$.

Some disaccharides, e.g. maltose, will reduce Benedict's (Fehlings) Reagent but others, e.g. sucrose, are non-reducing sugars. All disaccharides are sweet and soluble in water. One important disaccharide, sucrose, is transported in the sieve elements of the phloem and occasionally

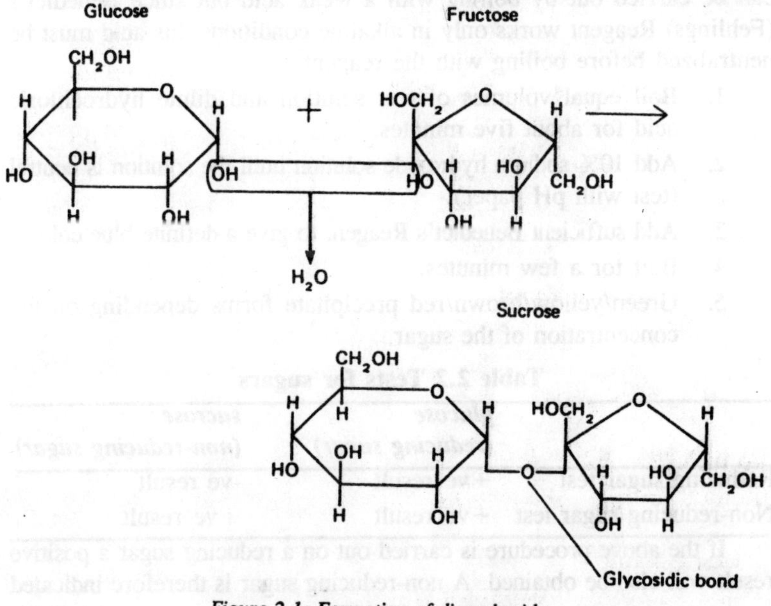

Figure 2.1: Formation of disaccharide.

stored in parenchyma cells, as in the scale leaves of onion. Disaccharides are readily converted into respiratory substrates.

Polysaccharides 'many sugars'

These are formed by the condensation of numerous monosaccharides. e.g. starch; glycogen; cellulose

They have the general formula $(C_6H_{10}O_5)_n$

They will not reduce Benedict's (Fehlings) Reagent, are not sweet, are insoluble and are used for storage (starch; glycogen) or for structural support (cellulose).

Tests for carbohydrates

Reducing Sugars

Using a little of the carbohydrate (if necessary by grinding with a little water in a mortar and pestle), add an equal volume of Benedict's (Fehlings) Reagent and boil for a few minutes. Presence of a reducing sugar is indicated by a green/yellow/brown/red precipitate depending on the concentration of the sugar.

Non-reducing Sugars

A non-reducing sugar, e.g. sucrose, will not give a precipitate when heated directly with Benedict's (Fehlings) Reagent unless it is first hydrolysed into its component reducing sugars. This hydrolysis can be carried out by boiling with a weak acid but since Benedict's (Fehlings) Reagent works only in alkaline conditions this acid must be neutralized before boiling with the reagent.

1. Boil equal volumes of test solution and dilute hydrochloric acid for about five minutes.
2. Add 10% sodium hydroxide solution until the solution is neutral (test with pH paper).
3. Add sufficient Benedict's Reagent to give a definite blue colour.
4. Boil for a few minutes.
5. Green/yellow/brown/red precipitate forms depending on the concentration of the sugar.

Table 2.2 Tests for sugars

	glucose (reducing sugar)	***sucrose (non-reducing sugar)***
Reducing sugar test	+ve result	-ve result
Non-reducing sugar test	+ve result	+ve result

If the above procedure is carried out on a reducing sugar a positive result will also be obtained. A non-reducing sugar is therefore indicated by a negative result for a reducing sugar followed by a positive one for a non-reducing sugar.

Where both sucrose and glucose are present in a solution the following procedure should be used:

The reducing sugar test and the non-reducing sugar test are carried out separately on two equivalent samples and, provided equal volumes of Benedict's Reagent are used, the amount of precipitate after the non-reducing sugar test will be greater than after the reducing sugar test.

Starch

Add iodine in potassium iodide solution to the substance to be tested. Do not heat. The presence of starch is indicated by a blue-black coloration.

Lipids (Fats)

Lipids are formed as shown in Figure elsewhere in this chapter.

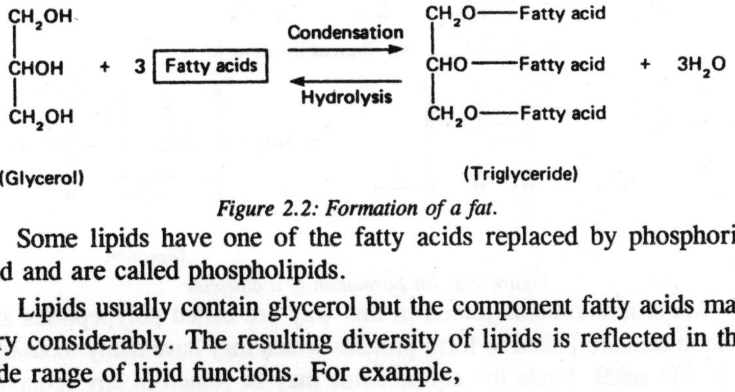

$$\begin{array}{l} CH_2OH \\ | \\ CHOH \\ | \\ CH_2OH \end{array} + 3\ \boxed{\text{Fatty acids}} \underset{\text{Hydrolysis}}{\overset{\text{Condensation}}{\rightleftharpoons}} \begin{array}{l} CH_2O\text{—Fatty acid} \\ | \\ CHO\text{—Fatty acid} \\ | \\ CH_2O\text{—Fatty acid} \end{array} + 3H_2O$$

(Glycerol) (Triglyceride)

Figure 2.2: Formation of a fat.

Some lipids have one of the fatty acids replaced by phosphoric acid and are called phospholipids.

Lipids usually contain glycerol but the component fatty acids may vary considerably. The resulting diversity of lipids is reflected in the wide range of lipid functions. For example,

1. **energy source:** 38 kJ/g.
2. **storage:** since lipids store over twice as much energy for the same mass as carbohydrates, they are used when it is important to keep weight to a minimum, e.g. in fruits and seeds.
3. **protection:** stored around delicate organs, e.g. kidneys.
4. **insulation:** especially useful where hair/feathers etc. are of little use, i.e. under water (e.g. blubber in whales).
5. **waterproofing:** leaf cuticles are made of waxes; secretion from sebaceous glands.

Tests for lipids

1. If a few drops of lipid are dissolved in ethanol and an equal

volume of water added, a cloudy white suspension is produced.

2. Lipids take up the red stain Sudan III readily. If a lipid and water are shaken with Sudan 111, only the lipid stains red.
3. Greasemark test. If a drop of lipid and water are placed on filter paper and left to dry, only the lipid leaves a residual mark when held up to the light.

Proteins

These consist of long chains of amino acids, the basic formula of which is shown in Figure 3.3a.

There are over twenty naturally occurring amino acids which differ in the composition of the R group.

Two amino acids may be linked together by condensation.

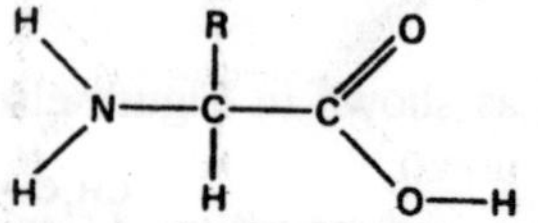

Figure 2.3: (a) Typical amino-acid

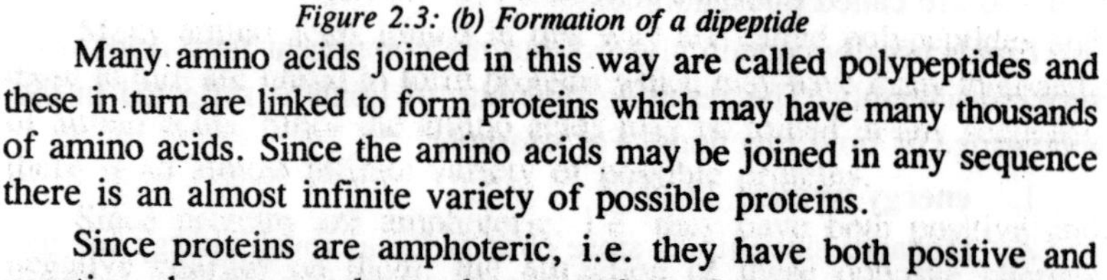

Figure 2.3: (b) Formation of a dipeptide

Many amino acids joined in this way are called polypeptides and these in turn are linked to form proteins which may have many thousands of amino acids. Since the amino acids may be joined in any sequence there is an almost infinite variety of possible proteins.

Since proteins are amphoteric, i.e. they have both positive and negative charges on them, the attraction of these opposite charges forms weak electrostatic (hydrogen) bonds causing the chain to form a complex three-dimensional structure-globular proteins.

All enzymes and some hormones are globular proteins and they also form antitoxins and antibodies. Sometimes the protein consists of long parallel chains with cross links -fibrous proteins. These are insoluble and have structural functions, e.g collagen in cartilage; keratin in hooves, feathers, hair; actin and myosin in muscle; in cell membranes.

If a globular protein is heated or treated with a strong acid or alkali the hydrogen bonds are broken and it reverts to a more fibrous nature-denaturation. Proteins sometimes occur in combination with a non-protein substance (prosthetic group); these are called conjugated proteins, e.g. haemoglobin.

Tests for proteins

1. To a quantity of test solution add half as much Millon's Reagent and boil. A red coloration indicates a protein. Since this is specific to the amino acid tyrosine it is possible that a few proteins (ones which lack tyrosine e.g. gelatin) will not give a positive result.
2. *Biuret Test*. To the test solution add an equal volume of 10% potassium hydroxide solution. Add 0.5% copper sulphate solution a drop at a time. A purple coloration indicates a protein. Care is needed since the addition of excess copper sulphate solution results in the disappearance of the coloration.

ENZYMES (INCLUDING BIOCHEMICAL TECHNIQUES)

Underlying Principles

The Laws of Thermodynamics

The First Law of Thermodynamics (Law of Conservation of Energy) states that energy may be transformed from one form into another but is neither created nor destroyed.

The Second Law of Thermodynamics states that processes involving energy transformations will not occur spontaneously unless there is a degradation of energy from a non-random to a random form.

Blackman's Law of Limiting Factors

states that when the speed of a process is conditioned by a number of separate factors, the rate of the process is limited by the pace of the slowest factor.

Reaction Kinetics

Chemical reactions are reversible, the direction being determined by the conditions in the reaction:

$$A \rightleftarrows B$$

The reaction may move from left to right at a high pH and in the reverse direction at a low pH. The addition of more A or the removal of B will tend to move the reaction from left to right and the addition of B or removal of A tends to move it from right to left.

Table 2.3: Enzyme Group Reactions.

Enzyme group	*Reaction catalysed*	*Example*
1. Oxidoreductases	All reactions of the oxidation-reduction type	Dehydrogenases; oxidases
2. Transferases	The transfer of a group from one substrate to another	Transaminases; phosphorylases
3. Hydrolases	Hydrolytic reactions	Phosphatases; peptidases; amidases; lipases
4. Lyases	Additions to a double bond or removal of a group from a substrate without hydrolysis, often leaving a compound containing a double bond	Decarboxylases
5. Isomerases	Reactions where the net result is an intramolecular rearrangement	All-trans retinal $\rightarrow$ all-cis retinal
6. Ligases	The formation of bonds between two substrate molecules using energy derived from the cleavage of a pyrophosphate bond such as ATP	Acetate + CoA-SH + ATP $\rightarrow$ acetyl-CoA + AMP + P-P

Points of Perspective

Historical background should include an awareness of Buchner who named enzymes, the word meaning 'in yeast'.

Candidates should consider medical, industrial and agricultural uses of enzymes, e.g. enzyme inhibitors as a basis of sulphonamide drugs and certain pesticides and the use of, enzymes in biological washing powders.

Essential Information

Names of enzymes are normally derived from the addition of the suffix -ase to the name of the substrate on which they act, e.g. lactase acts on lactose; maltase on maltose, etc. Enzymes are also grouped according to the type of reaction they carry out.

Characteristics of Enzymes

Specificity

The degree of specificity varies from those which will catalyse reactions of one isomer to those which catalyse reactions involving a particular type of linkage, e.g. peptide link, wherever it occurs.

Reversibility

Enzymes will catalyse reactions in either direction, e.g. carbonic anhydrase:

carbonic acid → carbon dioxide and water (in lungs)

$H_2CO_3 \rightarrow CO_2 + H_2O$

$CO_2 + H_2O \rightarrow H_2CO_3$ (in tissues)

The direction depends on conditions such as pH and substrate concentration.

Effect of enzyme concentration

The rate of an enzyme-catalysed reaction increases with enzyme concentration provided there is excess substrate.

Effect of substrate concentration

The rate of an enzyme-catalysed reaction increases as substrate concentration increases provided there is excess enzyme.

Temperature

Up to about 40°C the rate of an enzyme-catalysed reaction doubles for each 10°C rise in temperature; it then tails off until at about 60°C it stops. Temperature limits vary from enzyme to enzyme.

pH

Every enzyme has its own optimum pH at a given temperature,

concentration and substrate concentration. e.g.

pepsin	about pH 2.0
salivary amylase	about pH 7.0
arginase	about pH 10.0

Activation

Many enzymes require the presence of another substance before they will catalyse a reaction.

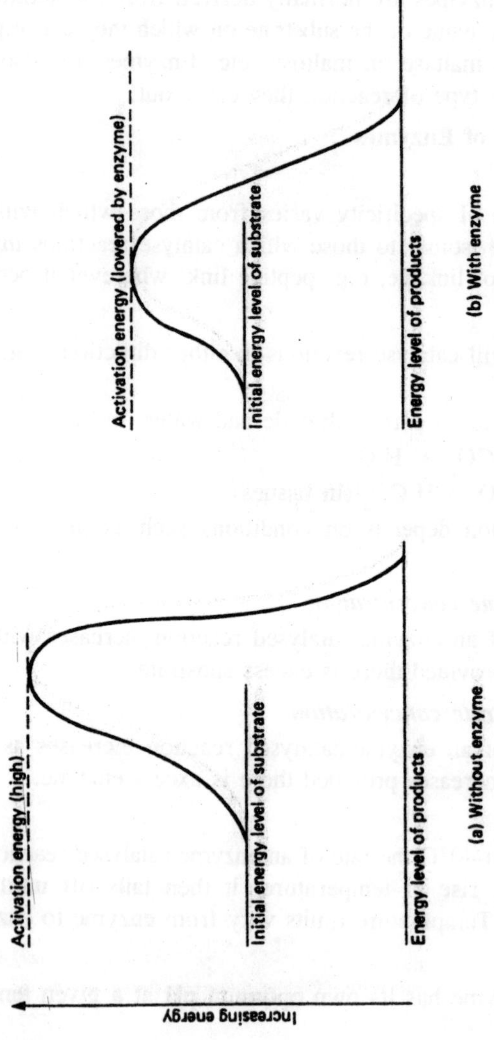

Figure 2.4: How enzymes lower activation energy levels.

Figure 2.5: The lock and key hypothesis

If such a substance is inorganic it is called a *co-factor*, e.g. calcium activates thrombokinase in the clotting of blood.

If it is a separate, non-protein organic molecule it is known as a *co-enzyme*, e.g. nicotinamide adenine dinucleotide (NAD) is a co-enzyme for dehydrogenases; the vitamin thiamine is a co-enzyme for pyruvic dehydrogenase.

If the non-protein, organic molecule forms an integral part of the enzyme it is called a *prosthetic group*, e.g. cytochrome oxidase has a prosthetic group containing iron.

Inhibition

Certain substances slow down or stop enzyme-controlled reactions. Where the inhibitor and substrate compete for the enzyme, neither forming a permanent combination, the inhibition is called competitive; the relative concentrations of substrate and inhibitor determine the degree of inhibition, e.g. malonic acid competes with succinic acid for succinic dehydrogenase.

In non-competitive inhibition, the inhibitor forms a permanent combination with the enzyme to the exclusion of the substrate; the degree of inhibition therefore depends on the concentration of inhibitor only, e.g. cyanide inhibits cytochrome oxidase thus arresting respiration.

Inhibition of enzymes is a normal part of the control of metabolic pathways (negative feedback).

Mechanism of enzyme action

In any spontaneous reaction the energy level of the products is lower than the energy level of the initial substrate. Before the reaction can occur, the substrate molecule must surmount an energy barrier known as the activation energy.

The presence of an enzyme lowers the activation energy. The active site is a precise region of the enzyme molecule whose shape is complementary to that of the substrate and into which the substrate fits.

In molecular terms this is thought to be by enzyme and substrate forming a complex in which the two molecules fit neatly together -an explanation known as the lock and key hypothesis.

Enzyme specificity is explained by the need for enzyme and substrate to fit together precisely. The molecular configuration of enzymes is determined by weak electrostatic (hydrogen) bonds which are broken by heat, acids and alkalis; hence their sensitivity to temperature and pH.

Biochemical Techniques

An individual cell may carry out up to 1000 chemical reactions, which must be carefully controlled. To assist control the conversion of one chemical to another takes place in a series of small steps known as a metabolic pathway. In addition the energy is released gradually rather than in a violent manner.

To determine the order of intermediates, a number of laboratory techniques are used:

1. the addition of an intermediate which should lead to an increase in all intermediates between it and the product.
2. the addition of an enzyme inhibitor which should cause an accumulation of the enzyme's substrate and the substances formed prior to it.

Centrifugation

Cells are broken down mechanically to form an homogenate suspension which is spun at a high speed causing the heavy structural components to separate out and form a sediment below a supernatant liquid. Particles of a different mass may be separated by centrifuging at different speeds (differential centrifugation).

Dialysis

Small molecules such as inorganic ions may be separated from larger ones such as protein by placing them within a membrane permeable only to the ions. The membrane is then placed in water into which the ions diffuse.

Chromatography

This is a means of separating out mixtures of chemicals by using their different sol ubilities in certain solvents. A concentrated spot of the solution to be separated is placed at one corner of a strip of paper.

This is dipped in a solvent which carries the chemicals dissolved in it up the paper depositing them, the least soluble first, the most soluble last. If necessary other chemicals may be used to make the spots visible., e.g. Ninhydrin colours amino acids. They are identified by their colour and R_f values:

$$R_f = \frac{\text{distance moved by spot}}{\text{distance moved by solvent}}$$

Spots may be further separated by running a different solvent at right angles to the first (two-way chromatography), or, instead of paper, a layer of suitable material, e.g. silica gel, may be used (thin layer chromatography).

Electrophoresis

This is similar to chromatography except that the spot is put in the centre of a strip of paper or the thin layer soaked in a suitable electrolyte. A potential difference is applied across it so that cations move towards the anode and anions to the cathode separating the substances. not on their solubilities, but on their relative electropositivity and negativity.

Isotopes

Since an isotope differs from the normal element its progress along a metabolic pathway can be traced in a number of ways. If radioactive, e.g. ^{14}C, its presence in intermediates can be determined using Geiger counters once all the intermediates have been separated by one of the above methods.

If too little radiation is emitted, the chromatogram may be placed next to a photographic plate which becomes exposed by those spots containing the radioactive material. The developed photograph is called an autoradiogram.

If the isotopes are not radioactive they are detected and measured using a mass spectrometer which separates them according to their atomic mass.

X-ray diffraction

When a beam of X-rays is fired at a sample of the substance being studied, the atoms scatter the rays and the diffraction pattern so formed is recorded on a photographic plate. Rotation of the sample to allow patterns from different angles to be made gives a complete analysis.

Computers are used to determine the precise positions of atoms from the patterns. This is used, for example, in the determining of DNA and protein structure.

Colorimetry

Many substances are detected by the production of a colour on the addition of an indicator or test reagent. The intensity of this colour is usually proportional to the amount of the substance present, e.g. the concentration of a starch solution may be determined by the intensity of the blue produced on the addition of iodine in potassium iodide solution.

A colorimeter is an instrument which passes a beam of light through a sample of the solution under test (which is usually contained in a special flat-sided tube called a cuvette). The amount of light passing through alters the electrical resistance of a photocell, producing a deflection on the meter that measures the voltage across the cell.

The scale shows either optical density and/or percentage transmission of light. The colorimeter has many applications such as the measurement of the rate of starch hydrolysis by amylase over a period of time. It is necessary to draw a calibration graph of colorimeter readings against known concentrations of starch, from which the actual starch concentration for any given colorimeter reading can be found.

GENETIC CODE

Underlying Principles

The nature of an organism is determined by the arrangement of its cells, whose nature is in turn determined by the chemical processes within them. These processes are controlled by enzymes, all of which are proteins.

Any cell produces a wide variety of enzymes, differing from cell to cell and organism to organism. As the catalytic properties of these enzymes depend on the precise sequence of their amino acids, the process by which they are made must be accurate and, as enzymes are only made when required, carefully regulated.

The vast set of instructions for the production of these enzymes must be contained within the minute amount of genetic material which passes from generation to generation.

Points of Perspective

Details of the elucidation of DNA structure by Watson and Crick using X-ray diffraction, etc. The Jacob-Monod theory of gene action in controlling protein synthesis.

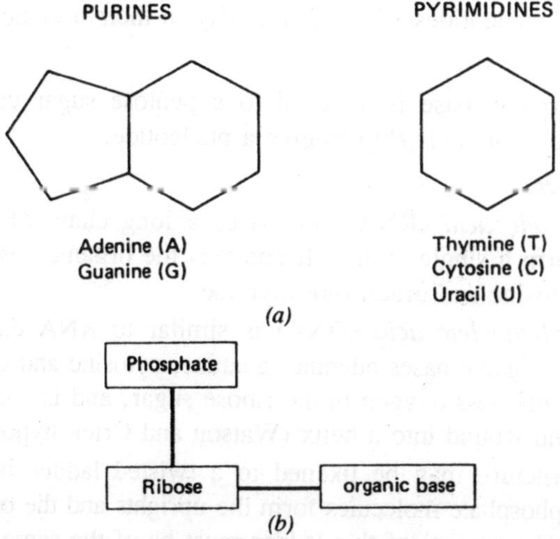

Figure 2.6: Basic structure of a nucleotide

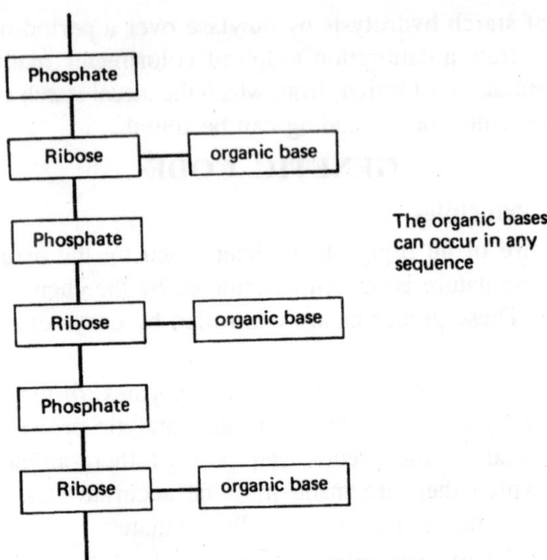

Figure 2.7: Portion of an RNA molecule

The Meselsohn-Stahl experiment using ^{15}N to demonstrate semi-conservative replication of DNA.

Hammerling's experiment on *Acetabularia* to show that the nucleus is involved in heredity.

Structure of Nucleic Acids

Nucleic acids are large molecules of high molecular mass composed of chains of nucleotides which differ only in their organic bases.

Nucleotides

Each organic base is attached to a pentose sugar called ribose and phosphoric acid (H_3PO_4) to give a nucleotide.

Nucleic acids

Ribonucleic acid (RNA) comprises a long chain of nucleotides linked to form a single strand. It contains the organic bases adenine, guanine, cytosine and uracil (not thymine).

Deoxyribonucleic acid (DNA) is similar to RNA except that it contains the organic bases adenine, guanine, cytosine and thymine (not uracil), has one less oxygen in the ribose sugar, and is composed of a double strand wound into a helix (Watson and Crick hypothesis).

The structure may be likened to a twisted ladder in which the ribose and phosphate molecules form the uprights and the organic bases the rungs. The 'rungs' of this ladder must be of the same length so it

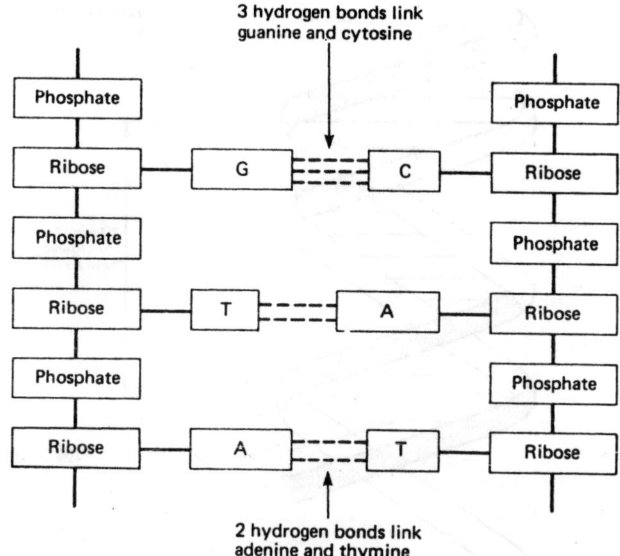

Figure 2.8: Basic DNA Structure-helix unwound.

follows that, as both purines and pyrimidines are found in DNA, each 'rung' must be made up of a purine linked with a pyrimidine. Two purines would make the 'rung' too long; two pyrimidines too short (see basic molecular shape).

Table 2.4: Differences in structure between DNA and RNA.

DNA	RNA
Organic bases are adenine, thymine,	Organic bases are adenine, uracil, guanine, cytosine guani-ne, cytosine
Double helix	Single strand
Deoxyribose (H)	Ribose (OH)
Mostly nuclear	Throughout the cell
More stable	Less stable
Permanent	Temporary
Insoluble	Soluble
One basic type	Three types: messenger, transfer, ribosomal
Concentration constant	Concentration varies according to cell type
Adenine, thymine/cytosine,	Adenine, uracil/cytosine,

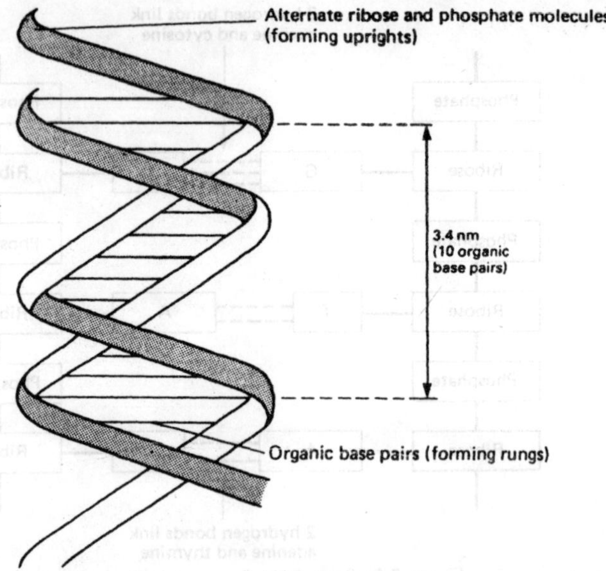

Figure 2.9: Double helical structure of DNA.

(Table Contd.)

guanine ratio	guanine ratio more variable about equal
Very large molecular mass (100 000-120 000 000)	Much smaller molecular mass (20 000-2 000 000)

Analysis of DNA shows the quantity of adenine and thymine to be the same and the quantity of guanine and cytosine to be the same, indicating that they form the organic base pairs.

Nucleic Acids and Protein Synthesis

All cellular substances are formed by chemical reactions involving enzymes, all of which are proteins. Thus protein synthesis is a fundamental process of life. There are two stages:

1. synthesis of amino acids.
2. assembly of amino acids in the correct sequence to form a particular protein.

Synthesis of amino acids

In plants, synthesis occurs in the mitochondria and chloroplasts in the following stages:

1. abstraction of nitrates from the soil

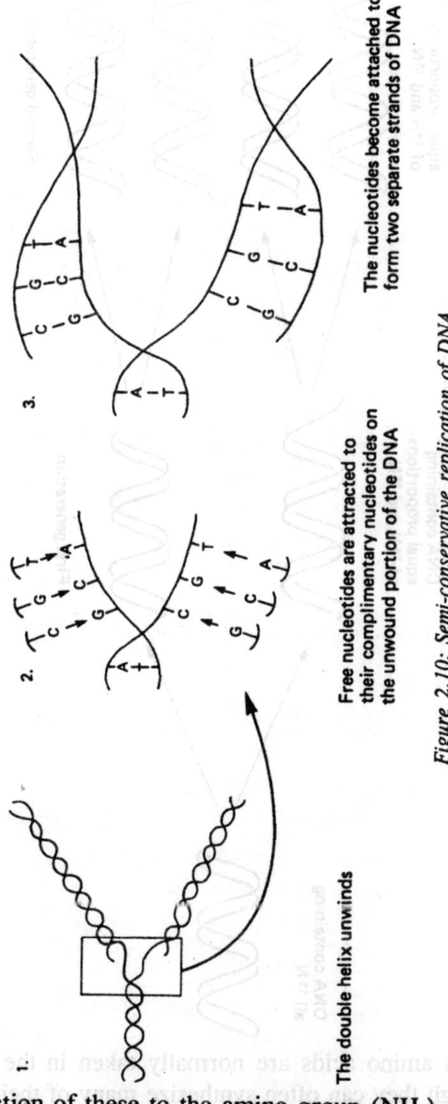

Figure 2.10: Semi-conservative replication of DNA.

2. reduction of these to the amino group (NH_2)
3. combination of amino group with a carbohydrate skeleton, e.g. a ketoglutaric acid from the Krebs' cycle.
4. transference of the amino groups from one carbohydrate skeleton to another *transamination* to form the twenty or more naturally occurring amino acids.

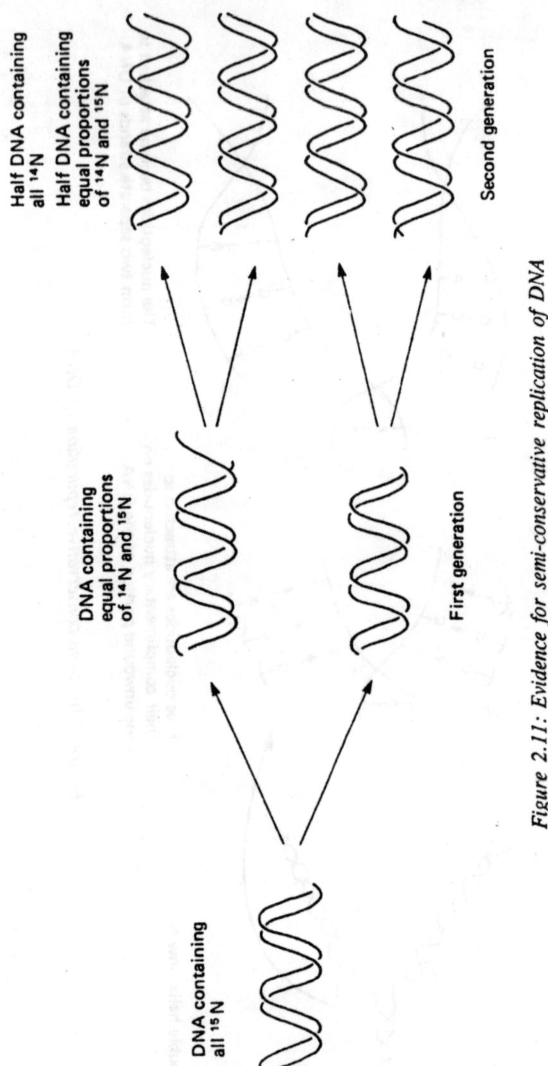

Figure 2.11: Evidence for semi-conservative replication of DNA

In animals amino acids are normally taken in the diet and used directly although they can often synthesize many of their own. In man all but nine can be synthesized; these are known as non-essential amino acids. The nine which cannot be synthesized - essential amino acids - must be provided in the diet.

Assembly of amino acids

The double helix of the DNA in the nucleus may unwind, leaving the organic bases unpaired; each of these may then attract its complementary nucleotide partner from free nucleotides within the nucleus to form two separate strands of DNA = **semi-conservative replication.**

Evidence for Semi-conservative DNA Replication

This was provided by the work of Meselsohn and Stahl in the late 1950s. Many generations of *E. coli* cells were grown in cultures containing ^{15}N (heavy nitrogen) to ensure that their DNA contained ^{15} N. They were then transferred to a medium containing only ^{14}N (light nitrogen).

The next two generations of *E. coli* each had their DNA separated by centrifugation and the presence of light and heavy nitrogen in the DNA was detected by its absorption of ultra violet light. The results showed the first generation of bacteria to have DNA containing equal proportions of ^{14}N and ^{15} N.

In the second generation half the bacteria had DNA containing all ^{14}N, the remainder having DNA with equal proportions of ^{14}N and ^{15} N. Sometimes the DNA only partially unwinds and the new strands produced become detached and leave the nucleus; these molecules are called messenger RNA (mRNA).

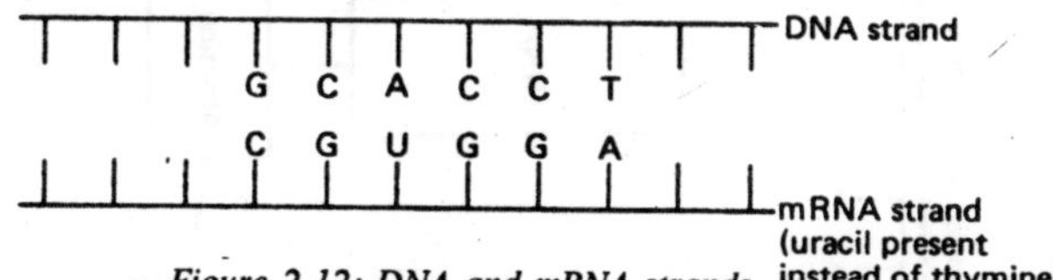

Figure 2.12: DNA and mRNA strands.

The mRNA has a definite sequence of organic bases along it which are complementary to the DNA from which it came. The mRNA moves out of the nucleus and wraps itself around groups of 5-50 ribosomes (polysomes) in the cytoplasm. In the cytoplasm are smaller strands of RNA called transfer RNA (tRNA) each of which links up with a specific amino acid at one end.

At the other end are three organic bases known as *anticodons*. When the mRNA wraps itself around the polysome, one of the ribosomes moves along it; as it passes each set of three bases (codons), the appropriate anticodon of the tRNA attaches.

As each codon on the mRNA consists of three bases, it is called a triplet code. Any amino acids may have up to six different triplet codes, e.g. in the diagram arginine is coded by CGU but it may also be coded for by CGC, CGA, CGG, AGA and AGG.

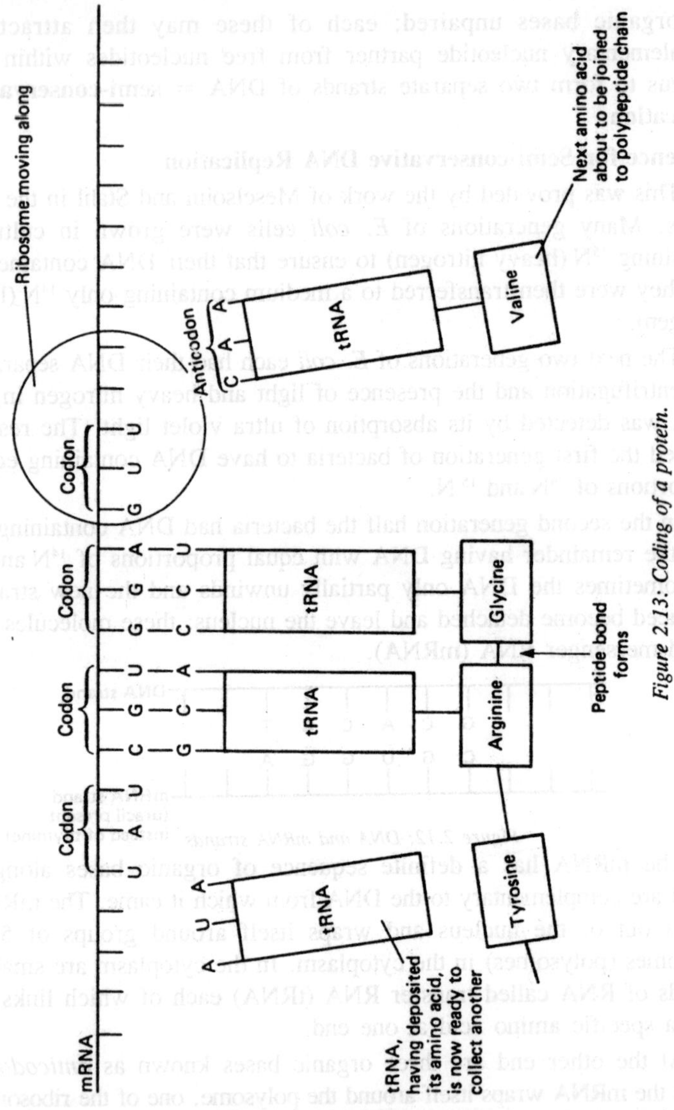

Figure 2.13: Coding of a protein.

A few codons, e.g. UAA; UAG; UGA, called *nonsense triplets*, do not specify amino acids but are used to indicate to the ribosome the end of a code. When the ribosome reaches the end of the mRNA molecule or a nonsense triplet, the chain of amino acids (polypeptides) becomes detached.

Ribosomes pass along the mRNA in rapid succession so that a

single strand of mRNA may produce large amounts of the same polypeptide chain. The region of DNA that codes for the production of a single polypeptide is called a *gene* (one gene-one polypeptide hypothesis). Because a protein may comprise more than one polypeptide, its formation may require more than a single gene.

Experimental evidence to show that DNA is the hereditary material

In *Pneumococcus,* the bacterium causing pneumonia, there are two easily distinguished strains: *avirulent*, which has no capsule (coat) and produces rough colonies (R) when grown on an agar plate; *virulent*, which has a polysaccharide capsule and produces smooth colonies (S). Smooth strains mutate into rough strains rarely but the reverse never occurs.

In 1928 Griffith injected mice with dead virulent (S) cells and living avirulent (R) cells. The mice frequently died of pneumonia and living virulent (S) cells were isolated from them. Thus the ability to cause the disease is somehow passed from the dead virulent (S) types to the living avirulent (R) types.

In 1933 Alloway showed that a cell-free extract was capable of bringing about this change; in 1944, by testing each component of the extract in turn, Avery, Macleod and McCarty showed that only a highly purified DNA extract was capable of bringing about this transformation.

Since the virulence passed on by the DNA is inherited in successive generations, DNA must be the hereditary material. In the 1950s Hershey and Chase, by radioactive labelling of the DNA in bacteriophages that attack *E, coli,* showed that only DNA entered the *E; coli;* as new phage particles were produced, it could only be the DNA which possessed the hereditary information necessary for the production of new viruses.

If the process is stopped halfway, i.e. only half the DNA is passed into *E.coli,* the construction of new phages is incomplete. Damage to DNA (e.g. change in nitrogen base) produces inherited changes (mutations).

When illuminated by varying wavelengths of ultraviolet light, the maximum absorption of DNA is of wavelength 254 nm, the wavelength shown to have the maximum mutagenic effect.

3

UNIT OF LIFE

THE CELL

Underlying Principles

Modern cell theory states that:

1. all living matter is composed of cells.
2. all cells arise from other cells.
3. all metabolic reactions of an organism take place in cells.
4. cells contain the hereditary information of the organisms of which they are a part, and this is passed from parent to daughter cell.

All cells are similar in comprising a self-contained and more or less self-sufficient unit surrounded by a cell membrane and having a nucleus at some stage of their existence.

At the same time cells show remarkable diversity of structure and function. In shape cells are basically spherical, although modification to suit function leads to a degree of diversity. In size they mostly range from 10-30μm in diameter. Their size is restricted by:

1. the surface area to volume ratio, which must be large to allow exchange of metabolic substances
2. the capacity of the nucleus to exercise control over the rest of the cell.

Points of Perspective

Historical background including the discovery of cells by Robert Hooke in 1665 and the cell theory by Theodore Schwann and Matthias

Schleiden in 1839. Electron microscopy (EM), including transmission EM and scanning EM, shauld be understood. Basic details of section-cutting, fixing and staining for EM preparations would be helpful.

A wide range of photoelectron micrographs of both plant and animal cells should be studied and practice obtained in identifying the parts and in calculating magnifications.

Essential Information

Cells were first seen by Hooke in 1665. Schleiden and Schwann in 1839 stated that all living systems are composed of cells and cell products (Cell Theory). Cells are studied using light or optical microscopes which have a resolution of approximately 0.25μm, imposed by the wavelength of light.

The electron microscope uses an electron beam instead of light and electromagnets instead of glass lenses. Electrons have a wavelength of 0.005 nm and in theory can resolve objects of 0.0025 nm in diameter, although in practice this is limited to 1 nm, i. e. an electron microscope has a resolving power approximately 400 $\times$ as great as a light microscope.

Disadvantages of the electron microscope:

1. preliminary treatment of material may cause distortion
2. specimens must be mounted in a vacuum and therefore be dead
3. expensive to make and run
4. only provides black and white image
5. preparation is lengthy and difficult
6. large, and needs a special room
7. affected by magnetic objects.

Cell Organelles

Plasma membrane

This separates the cell from its surroundings and has the same basic structure—known as the unit membrane—as the membranes of mitochondria, endoplasmic reticulum and other organelles.

There are a number of proposed models for the arrangement of protein and lipid in the unit membrane but most are variants of the Davson-Danielli hypothesis of the 1930s. In the early 1970s Singer and Nicholson proposed the *fluid-mosaic* model of the cell membrane based on chemical analyses, freeze-etching and electron microscopy. Their work confirms the bimolecular structure of the lipid part of the membrane

Differences between plant and animal cells:

Table 3.1: Plant and animal cell differences.

Plant	Animal
Cellulose cell wall as well as membrane	No cellulose cell wall, only membrane
Pits in cell wall	No pits
Plasmodesmata present	No plasmodesmata
Large vacuole filled with cell sap	Some vacuoles but usually small and numerous
Cytoplasm peripheral	Cytoplasm throughout the cell
Nucleus usually peripheral	Nucleus anywhere in cytoplasm but often central
2 cytoplasmic membranes: outer plasmalemma, inner tonoplast	Only 1 cytoplasmic membrane
Variety of plastids, e.g. chloroplast; leucoplast	Not normally plastids
Cilia and flagella absent in higher plants	Cilia common in higher animals
Centrioles absent in higher plants	Centrioles present

but indicates that the protein molecules form irregular globules and not a continuous layer.

Large particles enter the cell by phagocytosis: the cell membrane invaginates to form a depression containing the particles, and the invagination then becomes sealed off as a phagocytic vacuole. The process is selective, *e.g. Amoeba* engulfing food and phagocytes engulfing bacteria.

Pinocytosis is essentially similar to phagocytosis but involves the ingestion of liquids. In *Amoeba* pinocytic channels provide a means by which liquids are brought into the cell. Vacuoles form from the channels but the plasma membrane remains intact and liquids still have to cross it in order to enter the body of the cell.

Smaller, micropinocytic vacuoles have been seen through the electron microscope, for example at the base of microvilli in the mammalian small intestine.

Nucleus

This is bounded by a double unit membrane containing pores of 40-100 nm, allowing exchange of materials between the nucleus and the cytoplasm. The outer membrane is granulated and the inner one is

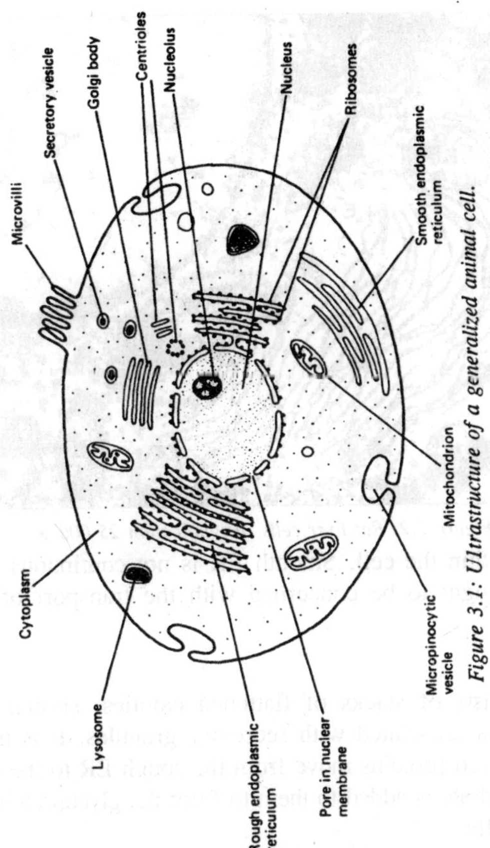

Figure 3.1: Ultrastructure of a generalized animal cell.

smooth. Inside the membrane is a meshwork of nucleoplasm interspersed with nuclear ribosomes and DNA. There may also be one or two small, round bodies, the nucleoli, which are rich in RNA and are not surrounded by a membrane.

Endoplasmic Reticulum

Throughout the matrix of the cell is a series of interconnected cavities bounded by unit membranes which are continuous with the nuclear membrane. This system of parallel cavities is known as the endoplasmic reticulum, or ER.

Most of the ER has granules called ribosomes attached to the matrix-side of the membranes. This is rough ER. Smooth ER lacks ribosomes and is thought to have a different function. Rough ER provides a means by which the polypeptides synthesized by the ribosomes are

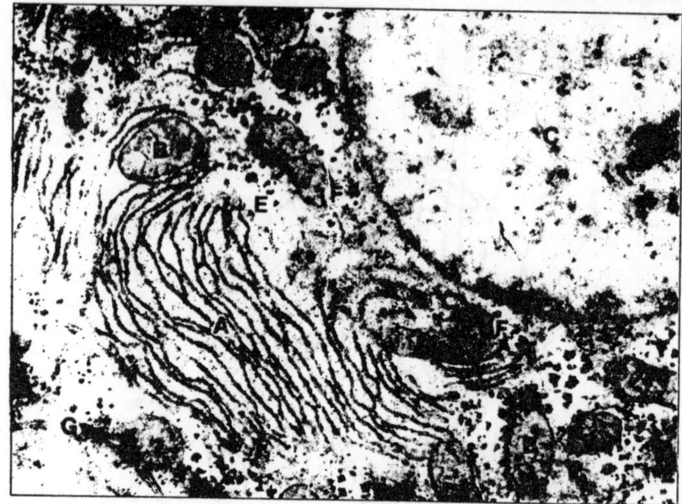

Figure 3.2: Rat liver cells, magnification 25 000 ×

transported within the cell. Smooth ER is not continuous with rough ER and is thought to be concerned with the transport of lipids and steroids.

Golgi Body

This consists of stacks of flattened cavities, similar to smooth ER, but always associated with secretory granules. It is thought that newly synthesized proteins move from the rough ER to the Golgi body where carbohydrate is added to them to form the glycoprotein secretions common in cells.

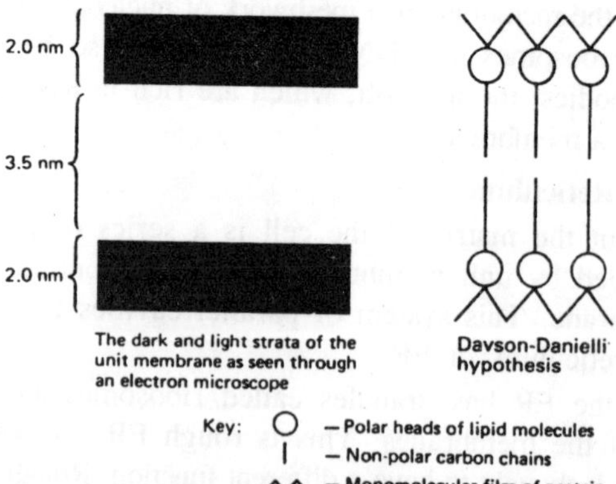

Figure 3.3: The organization of the unit membrane.

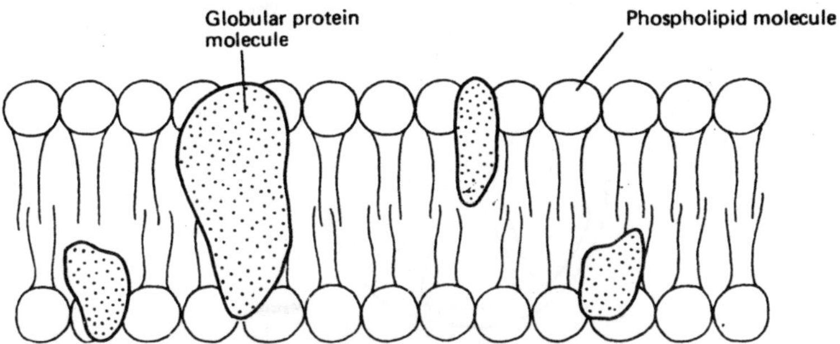

Figure 3.4· Fluid-mosaic model of membrane structure.

Part of the Golgi membrane surrounds these glycoproteins to form vesicles which move to the surface of the cell and discharge their contents.

Mitochondria

These vary in shape and size but are usually rod-shaped with a diameter of approximately 1.0 μm and length about 2.5 μm. The wall consists of two thin membranes, the inner one of which is folded to form cristae.

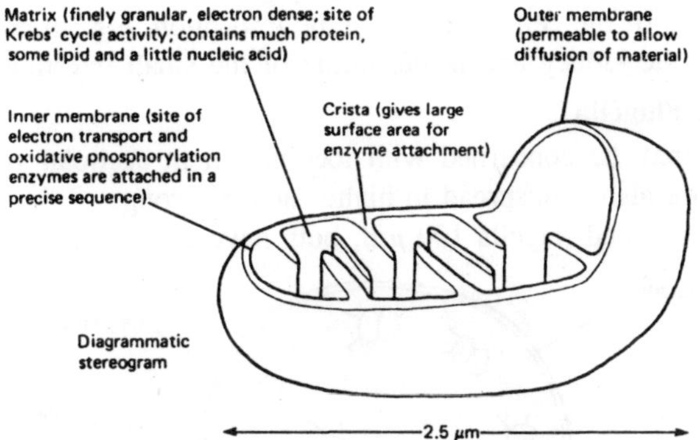

Figure 3.5: Structure of a mitochondrion.

Cell respiration takes place in mitochondria, most of the necessary enzymes being attached to the inner membrane and cristae, or occurring in the matrix. Thus cells which expend a lot of energy have numerous mitochondria each with many cristae, e.g. muscle cells.

Lysosomes

Sometimes called 'suicide-bags', these are spherical organelles containing enzymes capable of splitting complex substances into smaller

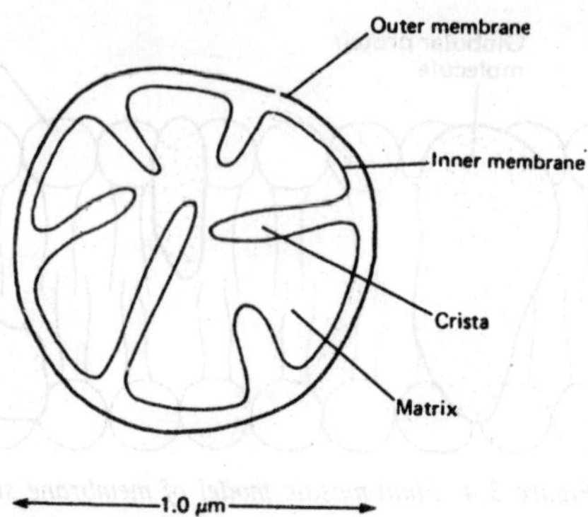

Figure 3.6: TS of a mitochondrion.

units. Their function is to destroy unwanted organelles (e.g. mitochondria), which are surrounded by a membrane into which the lysosomes discharge their enzymes. Lysosomes may also destroy entire 'worn-out' cells by liberating their enzymes within the cell.

Microvilli

These finger-like projections, approximately 1.0 μm long and 0.08 μm in diameter, on the surface of the plasma membrane increase the surface area of the cell, thereby aiding absorption, e.g. in convoluted tubules of the kidney and in the lining of the small intestine.

Cilia and Flagella

Both may be concerned with locomotion in acellular organisms but cilia are also widespread in higher animal groups. Cilia are about 5-10 μm long and flagella 100 μm; both have an average diameter of

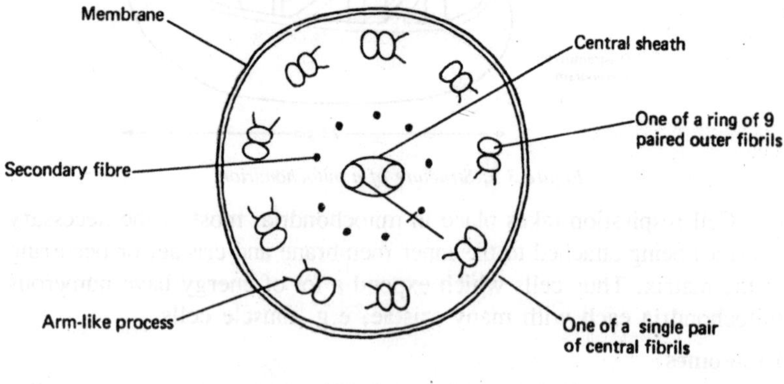

Figure 3.7: TS of a cilium, based on an electron micrograph.

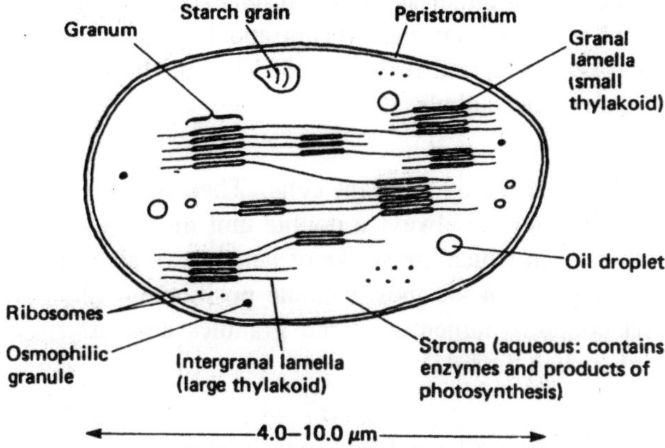

Figure 3.8: Structure of a chloroplast.

less than 0.3 μm. They have the same basic structure. At the base of each cilium is a basal body composed of nine peripheral fibres but not the central two. Movement of cilia and flagella requires ATP and the bending process is initiated at the base and transmitted towards the tip.

Centriole

In animal cells two are found at right angles to each other. Centrioles have the same basic structure as the basal bodies of cilia. They function in the formation of the spindle at cell division and also give rise to cilia and flagella.

Microtubules

These are made of protein and are widespread in the cytoplasm

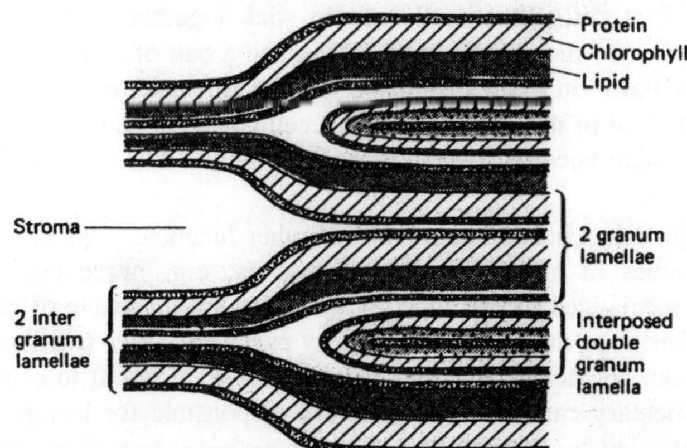

Figure 3.9: Section through grana showing arrangement of chlorophyll.

of the cell, occurring either singly or in bundles. They are thought to be concerned with cellular movements and transport in cells. They are easily broken down and reassembled. The spindle fibres formed at cell division are microtubules.

Chloroplasts

These are found only in plant cells. Their shape and size vary with species but there is always a double unit membrane surrounding a matrix (stroma) in which are stacks of lamellae (grana). The lamellae hold the chlorophyll in the most suitable position for photosynthesis. The stroma contains numerous starch granules and enzymes for the reduction of carbon dioxide.

Cell wall

This is a non-living layer found only in plant cells. A young plant cell has a primary cell wall composed of calcium pectate and some cellulose. As the cell ages a secondary cell wall is laid down; this is usually composed of layers of cellulose possibly impregnated with other substances, chiefly lignin.

In places the secondary cell wall is absent, giving rise to pits and the endoplasmic reticulum of adjacent cells may be continuous through the plasmodesmata.

Differentiation of Cells

Acellular organisms carry out all vital metabolic processes and have, within the cell, special organelles to carry out various functions. However, there is a limit to the degree of complexity possible within a single cell. To allow for increasing complexity the multicellular condition arose.

At first colonies of cells simply stick together, each cell being independent of the others. In the next stage a ball of cells is produced, but only those on the outside have flagella for movement. This marks the first stage of differentiation, i.e. a cell changing its normal structure so that it can become specialized to a particular function, in this case locomotion.

With differentiation comes loss of other functions, a process carried to extremes in higher plants and animals, e.g. nerve cells are so highly specialized to one function that they are incapable of dividing. Some functions must be carried out by every cell, e.g. respiration, but others can be 'delegated', e.g. digestion of food is left to cells lining the alimentary canal; muscle cells are responsible for locomotion.

Cells which carry out a particular function operate more efficiently when they are grouped together and this has led to the formation of a number of specialized tissues, e.g. epithelial; nervous; connective.

PLANT AND ANIMAL TISSUES

Underlying Principles

Fertilized egg cells divide to give a mass of similar (undifferentiated) cells. Although single cells can carry out all necessary functions, there are times when these conflict and it then becomes more efficient for each cell to take on a different function (differentiation) acquiring special characteristics to suit its function (specialization).

This inevitably leads to the loss of some functions and so to the dependence of one cell upon another. For further efficiency cells performing similar functions are grouped together to form a tissue. Some tissues contain very similar cells, e.g. epithelial tissue, and others are more diverse, e.g. connective tissue.

Points of Perspective

Embryological derivation of tissues from germ layers in animals. The derivation of plant tissues from meristems.

It is important to be able to correlate the structure of a tissue with its function.

The ability to interpret the three-dimensional nature of a tissue from longitudinal and transverse sections would be a considerable advantage. Students should be able to prepare temporary microscope slides of tissues, e.g. blood; leaf epidermis.

Essential Information

A tissue

This is a region consisting mainly of cells of a similar type and function bound together by cell walls (plants) or by intercellular material (animals).

Animal Tissues

The principal types and their derivations are: Epithelial -from all germ layers Connective - from mesoderm Muscular - from mesoderm Nervous - from ectoderm.

Epithelial Tissue

Associated with its role as a covering tissue subject to mechanical damage, epithelial tissue has the following characteristics:

1. cells frequently attached to basement membane
2. adjacent cells joined together by intercellular cement

Table 3.2: Functions of plant and animal tissues

Function	Plant	Animal
Protection	Epidermis	Skin
	Mechanical	Collenchyma Blood
	Sclerenchyma	Bone
	Cell sap of parenchyma	Cartilage
	Xylem fibres	Connective tissue proper
Secretion	Gland cells	Gland cells
Conduction of water or food	Xylem vessels	—
	Tracheids	
	Sieve tubes	
Conduction of impulses	—	Nerve
Contraction	—	Muscle

3. may be interconnecting bridges of cytoplasm between cells
4. may build up large number of layers to resist abrasion, e.g. epidermis of skin
5. rapid division of epithelial layer to resist abrasion and maintain only a thin layer of cells, e.g. intestines
6. resistance to abrasion may be increased by presence of keratin, e.g. epidermis of skin

Types of Epithelia

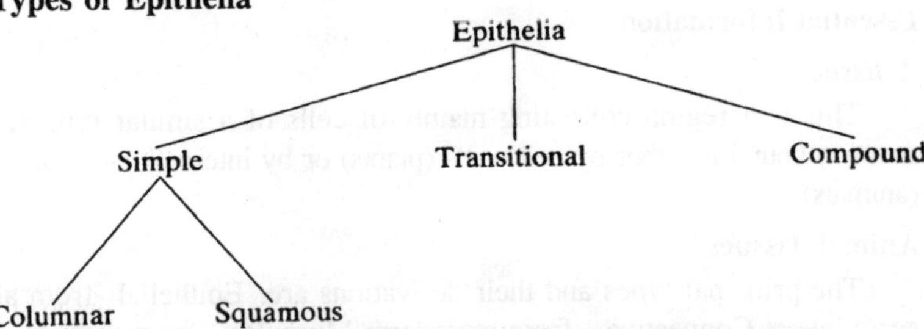

Any of the various types may be ciliated and/or secretory.

Columnar epithelium: cilia beat in a definite rhythm, moving along mucus and other particles. Found in, for example, nasal cavities; trachea; oviducts; ventricles of brain.

Squamous epithelium: sometimes called 'pavement' epithelium. Flattened, single layer of thin nucleated cells. Found where rapid diffusion is essential, e.g. Bowman's capsule; alveoli; endothelium.

Transitional epithelium: found where protection and distention

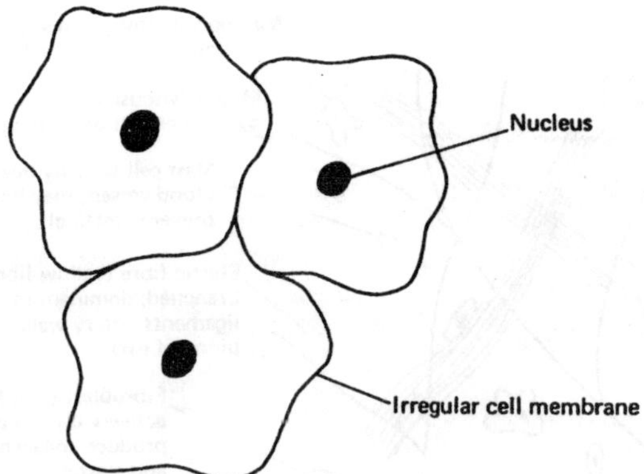

Figure 3.10: Squamous epithelium, surface view.

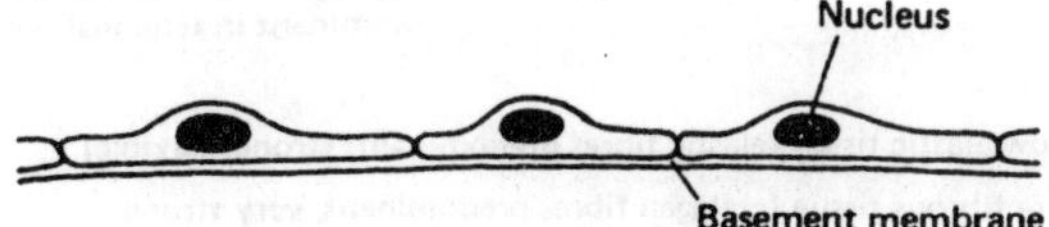

Figure 3.11: VS squamous epithelium

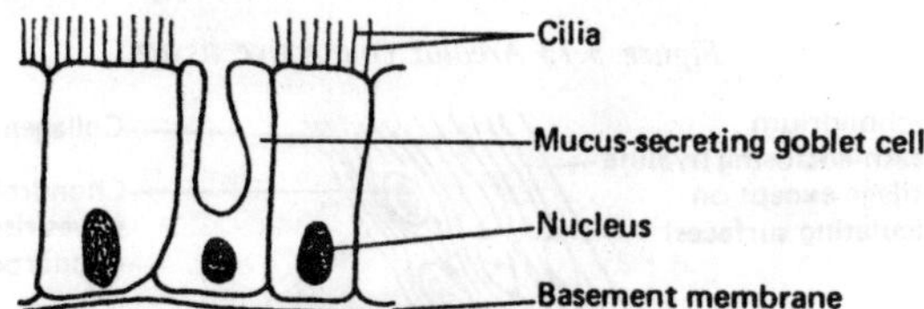

Figure 3.12: VS ciliated columnar epithelium.

are both required, e.g. bladder and urinary tract.

Compound (stratified) epithelium: many layers of cells; the ones nearest the basement membrane are usually columnar and those farthest from it are often flattened and dead, possibly impregnated with keratin. Found where there is considerable mechanical stress, e.g. epidermis of skin; oesophagus; anal canal; vagina.

Connective Tissue

Characteristics: variable appearance; includes a cellular element and the non-cellular product of these cells (the intercellular substance, or matrix).

Blood

Plasma is the matrix and the cells present include those shown in the table elsewhere in this chapter.

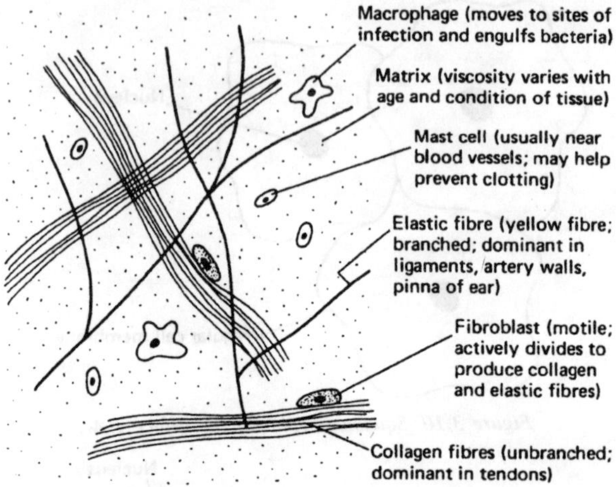

Yellow elastic tissue (elastic fibres predominant; strong; flexible)

White fibrous tissue (collagen fibres predominant; very strong; inextensible; limited flexibility)

Adipose tissue (matrix largely filled with fat cells; used for storage; especially under skin and around kidneys in mammals)

Figure 3.13 Areolar connective tissue.

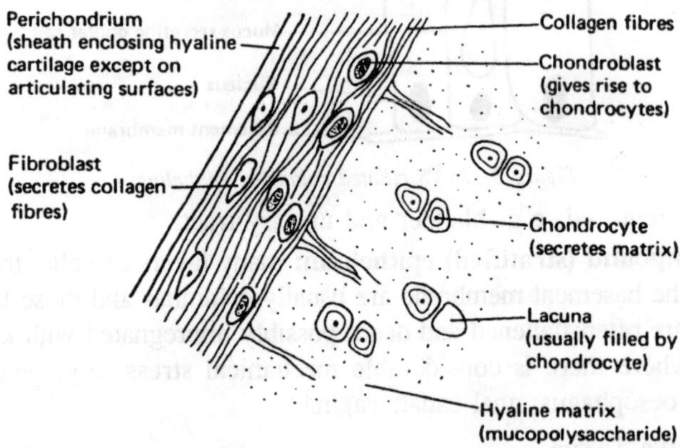

Figure 3.14: Hyaline cartilage.

Cartilage

Matrix is firm and resilient and secreted by chondroblasts. Proportions of fibres present vary in different types of cartilage. Hyaline cartilage is widespread in the embryo and persists in the adult in the trachea, larynx and articulating surfaces of bones. See in the otehr section, Locomotion and support in plants and animals.

Table 3.3 Frequency and functions of blood cells

Cell	No/mm³ in man	Functions
Erythrocyte	4.2-6.4 million	Carriage of oxygen as oxyhaemoglobin
Polymorphonuclear	1500-7500	Engulf bacteria leucocyte
Eosinophil	0-400	Produce antitoxins
Basophil	0-200	? but produce heparin and histamine
Lymphocyte	1000-4500	Produce antibodies
Monocyte	0-800	Engulf bacteria and coarse debris
Platelet (not truly cellular)	300 000	In coagulation

Types of connective tissue

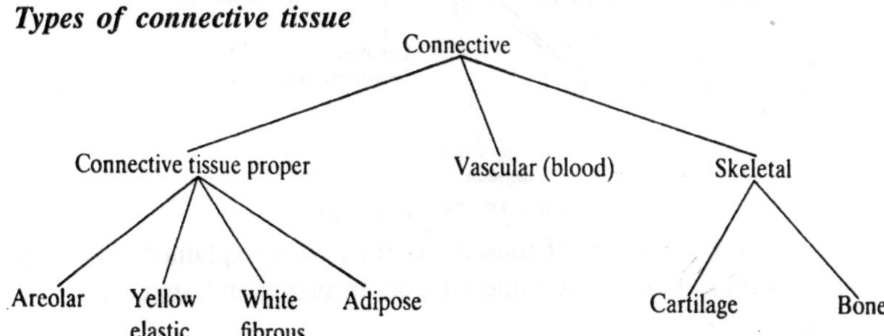

Bone

The five main functions of bone are:

1. framework for support, e.g. backbone
2. system of levers for movement
3. protection of vital organs, e.g. skull and ribs
4. reservoir for calcium and phosphorus 5 site of blood cell production in marrow.

Muscle

Characteristics

All muscle cells are contractile; cells bound together by connective tissue. Types: skeletal (striated; voluntary; striped); smooth (unstriated; involuntary; unstriped); cardiac.

Skeletal muscle

This forms the bulk of body muscle and is generally under voluntary

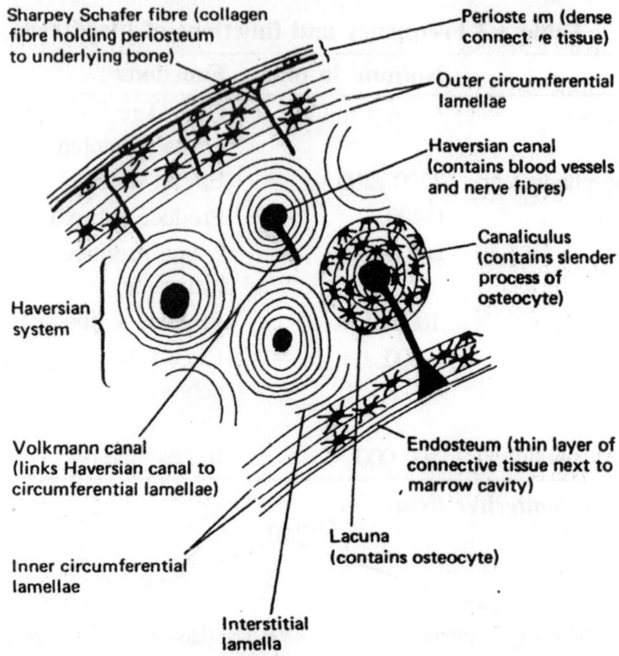

Figure 3.15: TS compact bone.

control. The mechanism of muscle contraction is explained elsewhere in this chapter, Locomotion and support in plants and animals.

Smooth muscle

This is often related to metabolic functions, e.g. nutrition; excretion. Smooth muscle does not contract very rapidly or powerfully but it never fatigues. It is found in the alimentary canal; skin; arteries (and to a lesser extent in veins); ducts; urinogenital tract.

Cardiac muscle

This is found only in the heart. It contracts powerfully and rapidly without fatigue.

Nervous Tissue

Characteristics: This has highly developed properties of irritability and conductivity. Nervous tissue is made up of neurones. Other types of neurones and the transmission of nerve impulses are dealt with in other section of this chapter, Nervous system and behaviour.

Parenchyma

These unspecialized cells are the major component of the ground

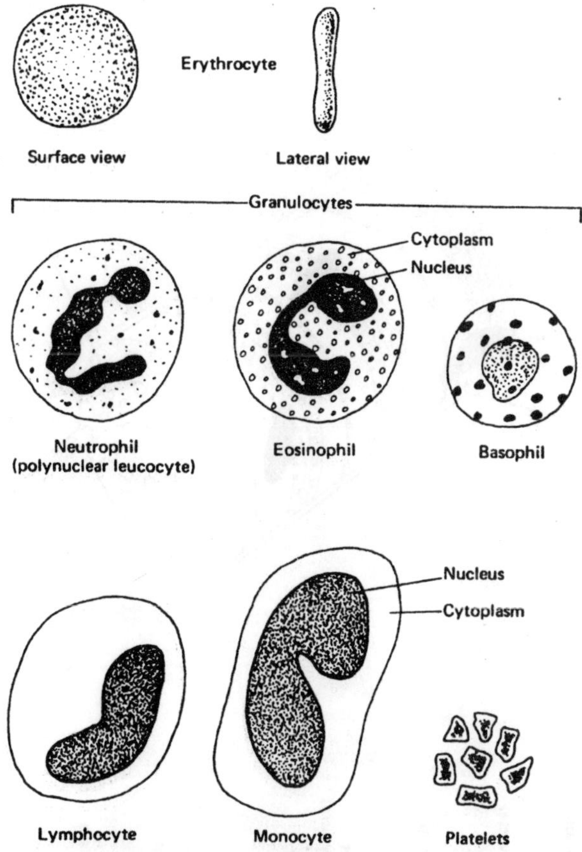

Figure 3.16: Types of blood cells.

tissue and are present in vascular tissues. They are potentially meristematic.

Collenchyma

The primary cell wall is unevenly thickened with cellulose and pectic substances but not lignin. Pits are present in the walls. Their function is support, especially in younger stems where plasticity is necessary to accommodate changing growth requirements.

Sclerenchyma

This is a supporting tissue with a secondary cell wall of lignin deposited on the primary cell wall of cellulose. The two basic types are sclereids and fibres but the differences between them are not always clear-cut. The pits in the cell walls may be simple or bordered.

Epimysium (continuous with tendon)
Perimysium
Blood vessel
Endomysin
Fibre (each is a syncytium)

Figure 3.17: (a) (i) TS whole skeletal muscle.

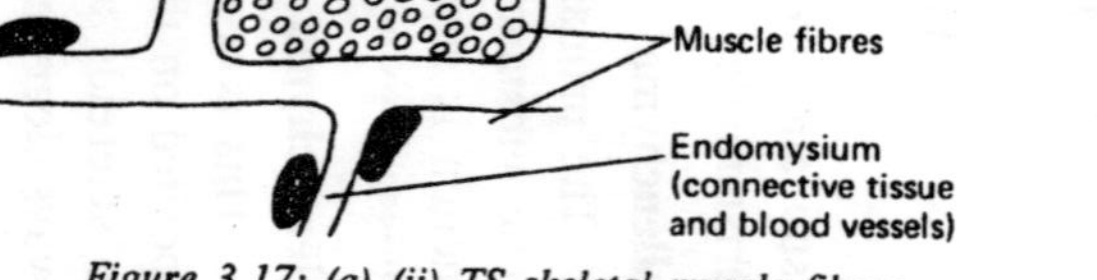

Figure 3.17: (a) (ii) TS skeletal muscle fibres.

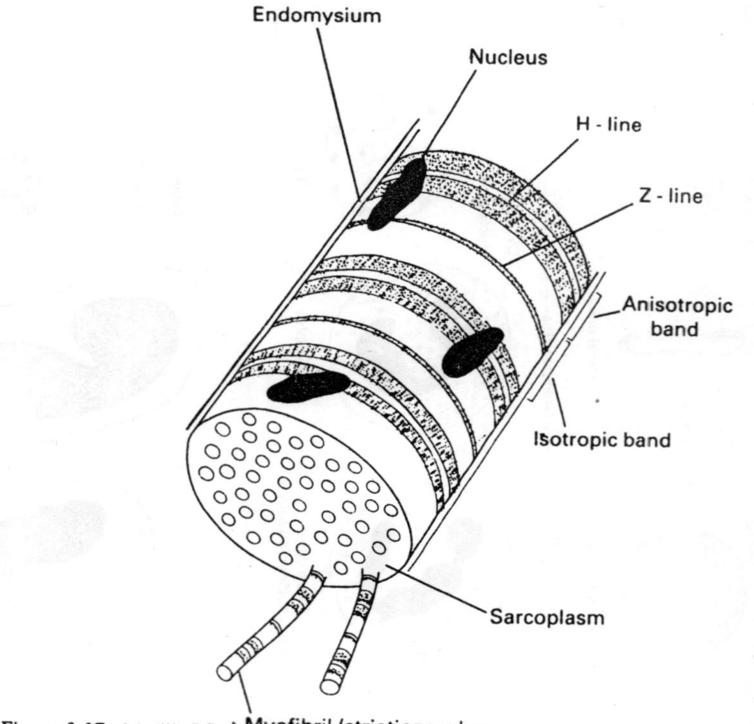

Figure 3.17: (a) (iii) LS skeletal muscle fibre.

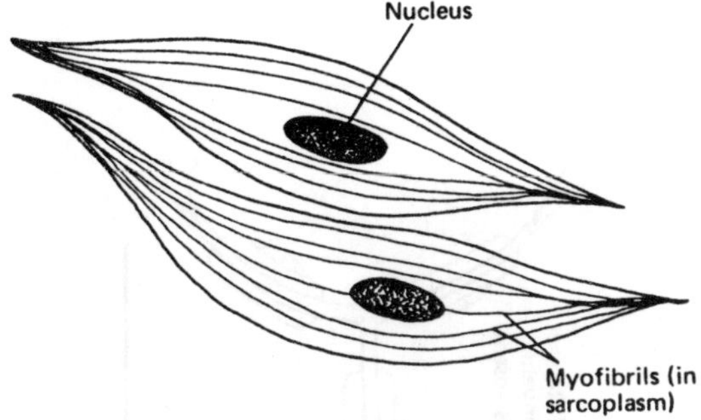

NB There are no cross striations

Figure 3.17: (b) LS smooth muscle fibres.

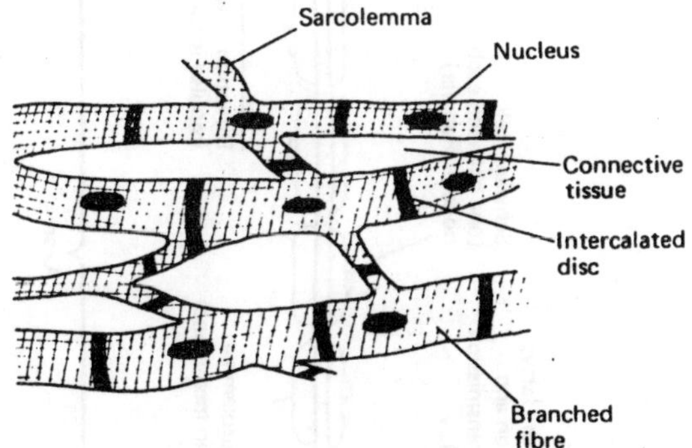

Figure 3.17: (c) Cardiac muscle.

Fibres

These are often found in the vascular bundles of dicotyledons or around the vascular bundles of monocotyledons. They are often grouped in strands. Jute, hemp and flax are economically important fibres.

Sclereids

These are widely distributed, heavily lignified cells with numerous pits.

Xylem

Primary xylem is formed from the embryo and the resultant meristems. Secondary xylem develops later, during secondary thickening. Xylem is made up of four main elements: vessels, tracheids, fibres and parenchyma.

Vessels: These cells have one or more perforations at each end so that water can move easily from cell to cell. Cell walls may be simple or perforated by bordered pits.

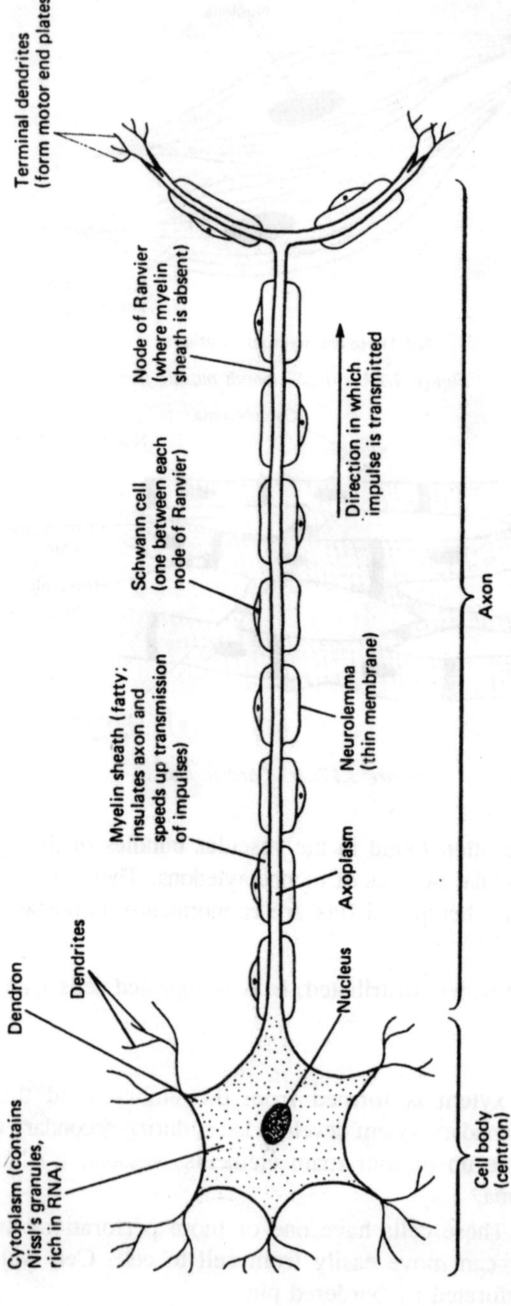

Figure 3.18: Typical motor neurone.

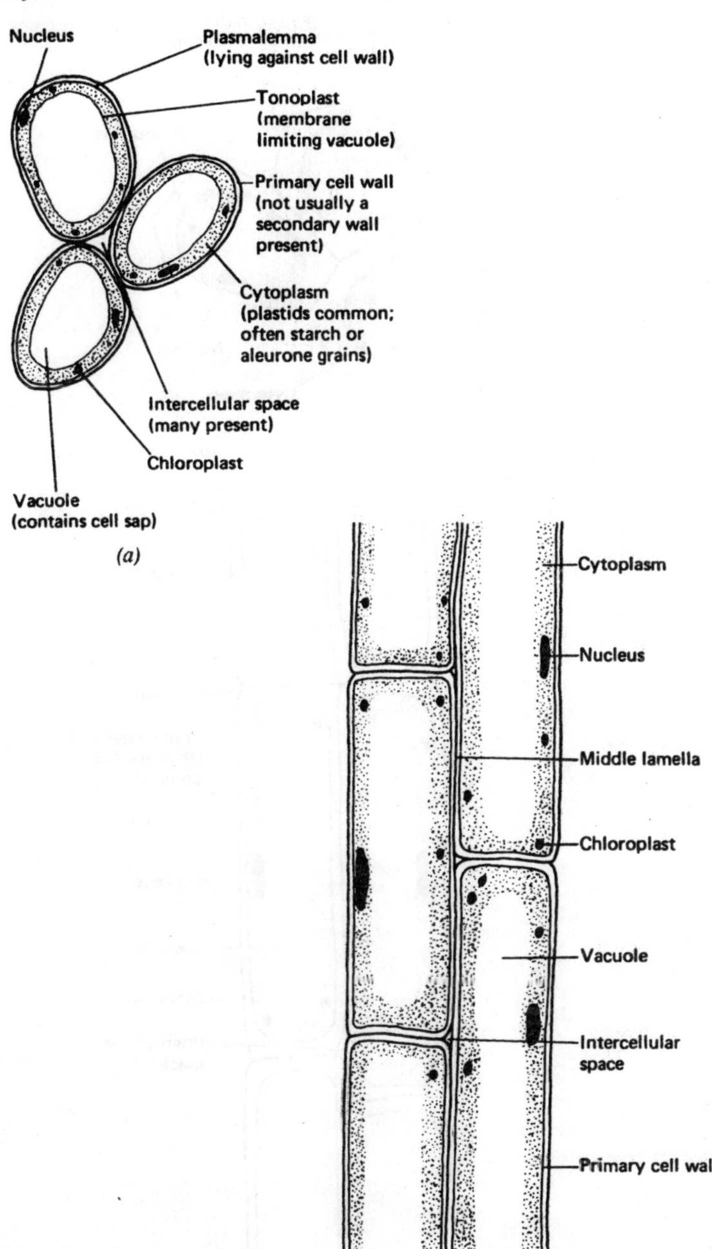

Figure 3.19: TS parenchyma

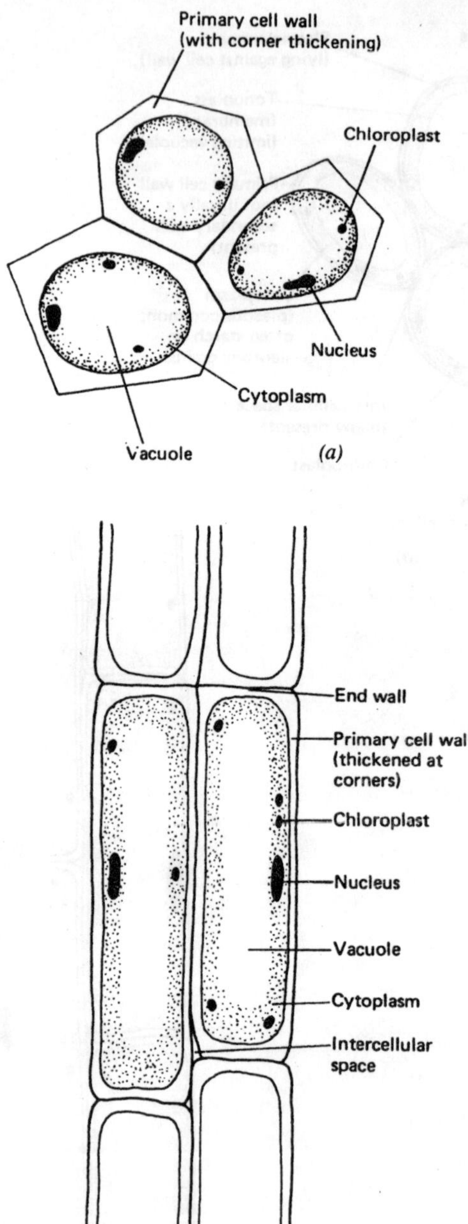

Figure 3.20: TS Collenchyma

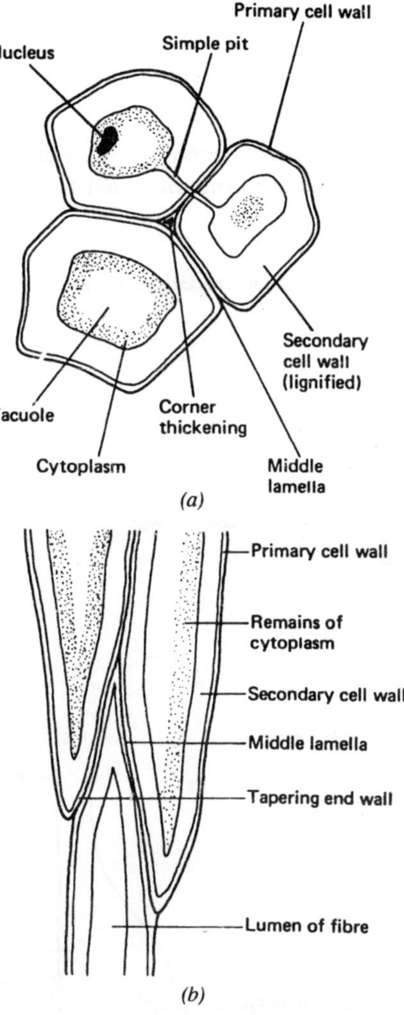

Figure 3.21: TS fibre.

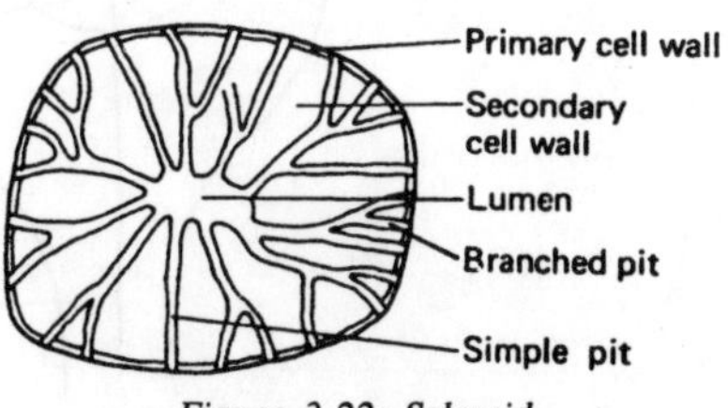

Figure 3.22: Sclereid.

Tracheids

These are imperforate cells but the pits occur in pairs so that the water can pass easily through the thin pit membrane.

Fibres

These are long cells whose secondary walls are commonly lignified. Pits are frequent. Fibres are primarily used for support.

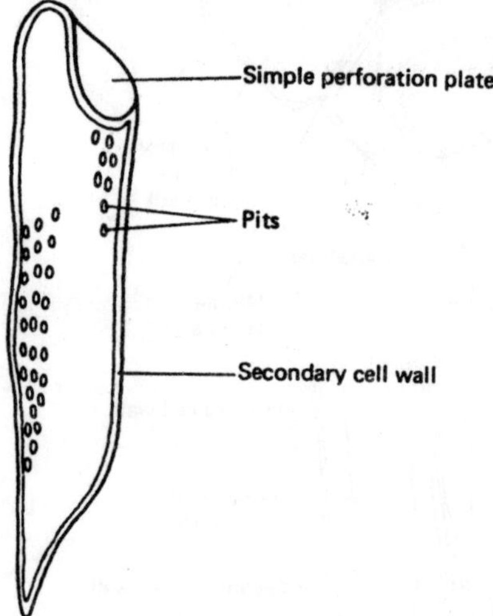

Figure 3.23: Vessel.

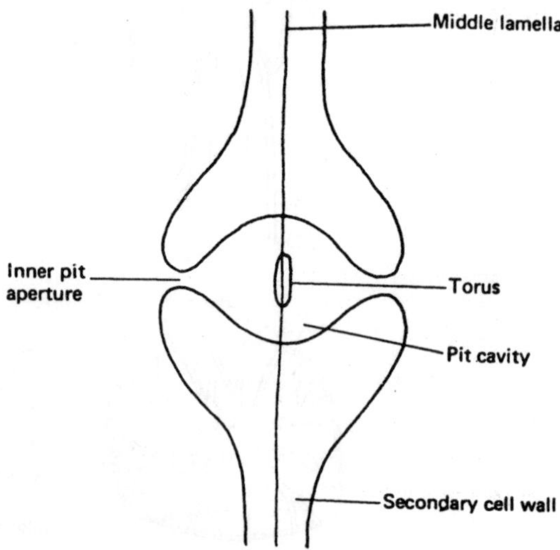

Figure 3.24: LS bordered pit (from racheid).

Parenchyma

These cells often store starch and oils.

In primary xylem the secondary walls which develop have very characteristic forms of thickening, e.g. annular; spiral; reticulate.

Phloem

Primary phloem develops from the procambium and secondary phloem from the vascular cambium. Phloem is made up of four main elements: sieve tubes, companion cells, sclerenchyma and parenchyma.

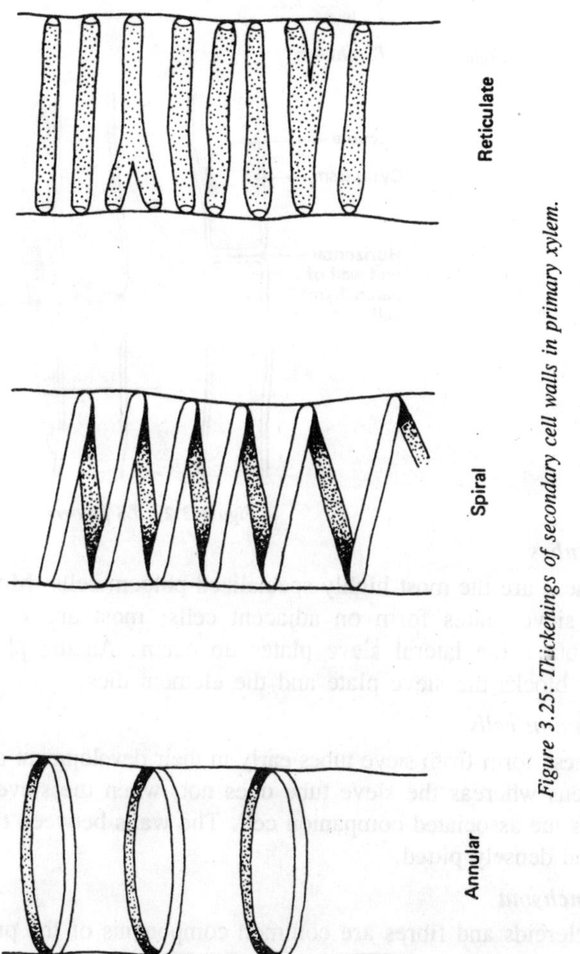

Figure 3.25: Thickenings of secondary cell walls in primary xylem.

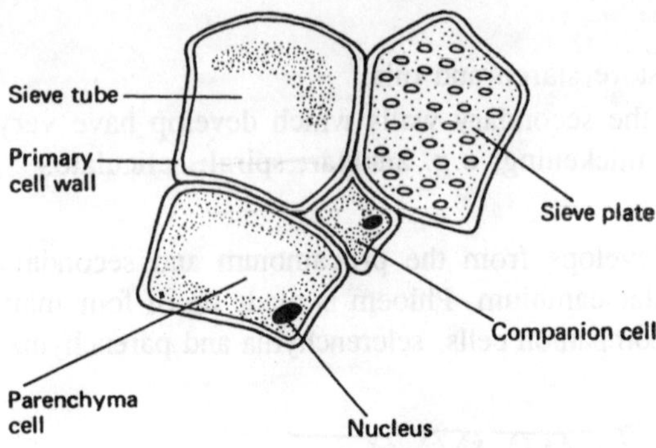

Figure 3.26: TS phloem.

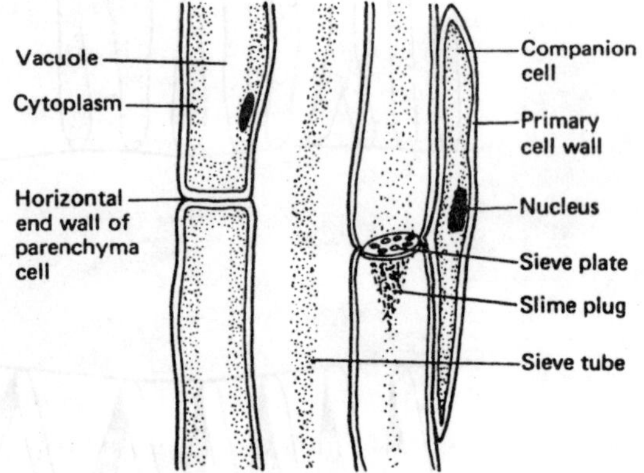

Figure 3.27: LS phloem.

Sieve tubes

These are the most highly specialized phloem cells. Modified pits called sieve plates form on adjacent cells; most are in a vertical series but some lateral sieve plates do occur. As the phloem ages callose blocks the sieve plate and the element dies.

Companion cells

These form from sieve tubes early in their development and contain a nucleus whereas the sieve tube does not; when the sieve tube dies so does the associated companion cell. The wails between the two are thin and densely pitted.

Sclerenchyma

Sclereids and fibres are common components of the primary and secondary phloem.

Parenchyma

These cells contain substances such as starch and tannins. Companion cells are specialized parenchyma cells.

4

Preservation and Inheritance

MITOSIS AND MEIOSIS

Underlying Principles

If a tissue is to be extended by growth, or damaged cells in it are to be replaced, it is important that the new cells are exact copies of the originals. In the same way a species that is successful colonizing a particular habitat may benefit from rapid multiplication to produce identical individuals which will also be suited to that environment.

In both cases a rapid form of cell division that produces identical daughter cells is needed. This is mitosis, which maintains the constancy of a species. The long-term survival of a species is, however, dependent on its ability to adapt to the constantly changing climatic, edaphic and biotic environment in which it lives, and on its ability to colonize new and different habitats.

To achieve this the offspring need to be different from their parents so that each has the potential to survive in a different environment and thus continue the species. A method of producing new characteristics by the mixing of genetic material within and between organisms has therefore evolved.

If the genes of two individuals are to be combined while maintaining the total chromosome number, the genetic material from each parent must be halved. This is meiosis, which creates variety within a species.

Points of Perspective

Both practical and theoretical examinations often contain questions involving identification from photographs of the stages of mitosis and meiosis. A careful study of photographs of all stages of both types of cell division is therefore invaluable.

Experimental details of how to obtain microscope preparations of the stages of meiosis and mitosis are often required (see question analysis). It is clearly an advantage if the candidate has performed these experiments.

Essential Information

Structure of chromosomes

Chromosomes are rod-like structures, consisting of nucleic acids and protein, located within the nucleus. During cell division chromosomes change their length as a result of coiling and uncoiling, dividing to form paired, joined *chromatids*.

Each chromosome has somewhere along its length a well-defined region where the chromatids are particularly closely associated and which seems to be the point at which force is exerted in the separation of dividing chromosomes.

This structure is called the *centromere*. Along the length of the chromosome there may be distinct constrictions and bumps, the pattern of which is quite constant for a particular chromosome from cell to cell. These 'bumps' are called *chromomeres* and they are probably caused by the coiling within the chromatids.

The number of chromosomes per nucleus is normally constant for all the individuals of a species, e.g. man has 46, rat 42. garden pea 14, tomato 24. The number of chromosomes characteristic of a species gives no indication of its level of organization.

Chromosomes are present in pairs and therefore it is often convenient to speak of the chromosome number of a particular species in terms of the number of pairs (i.e. 23 for man).

The members of each pair are alike but the different pairs are distinguishable. Every body (*somatic*) cell contains the characteristic number of chromosomes but mature germ cells (*gametes*) contain only half the usual number, one member of each pair. The gametes are described as *haploid* in chromosome number and the somatic cells as *diploid*.

Mitosis

In mitosis each chromosome duplicates itself and the duplicates

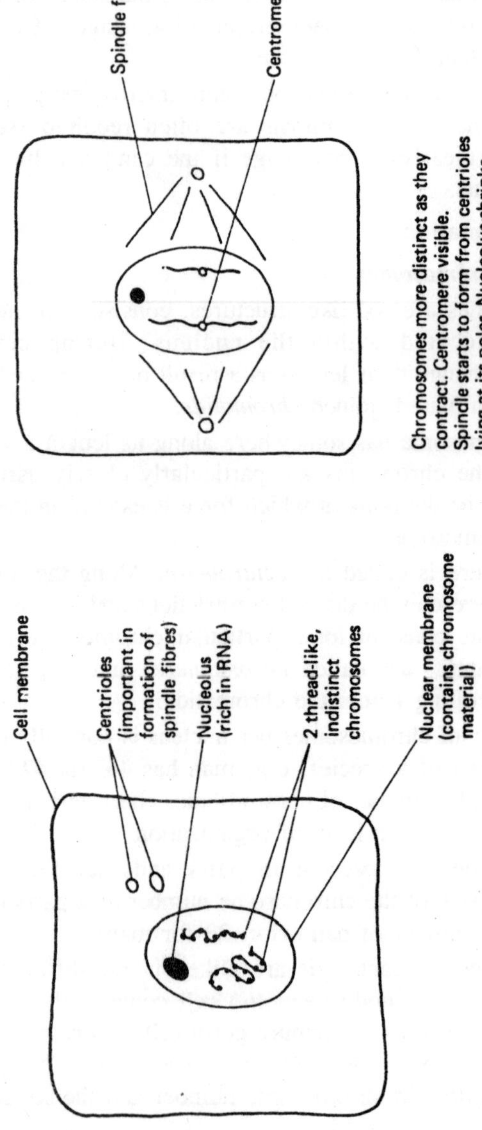

Figure 4.1: Stages of mitosis. (Figure Contd.)

(c) Late prophase

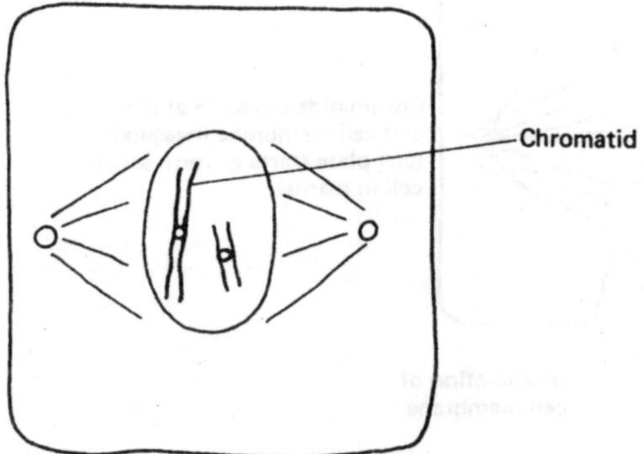

Chromatids visible as chromosomes shorten and thicken. Nucleolus disappeared. Nuclear membrane disappearing.

(d) Early metaphase

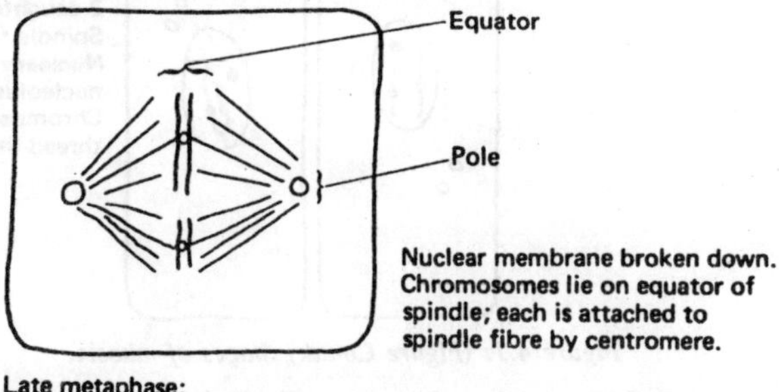

Nuclear membrane broken down. Chromosomes lie on equator of spindle; each is attached to spindle fibre by centromere.

Late metaphase: chromatids start to move apart.

(e) Early anaphase

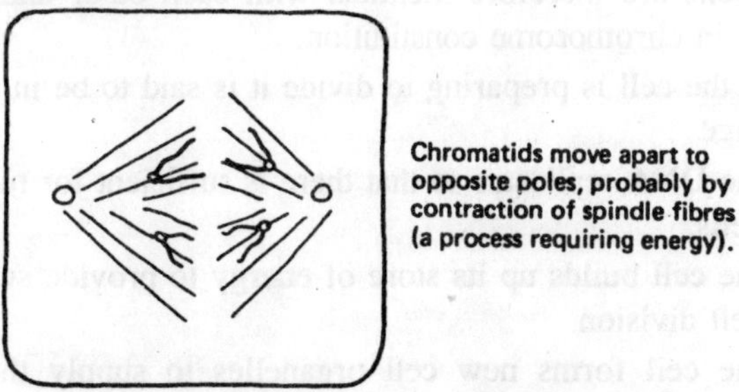

Chromatids move apart to opposite poles, probably by contraction of spindle fibres (a process requiring energy).

Late anaphase: chromatids reaching poles.

Figure 4.1: (Figure Contd.) Stages of mitosis.

(f) Early telophase

Chromatids assemble at poles and cell membrane invaginates (cell plate starts to form across cell in plants).

Invagination of cell membrane

(g) Late telophase

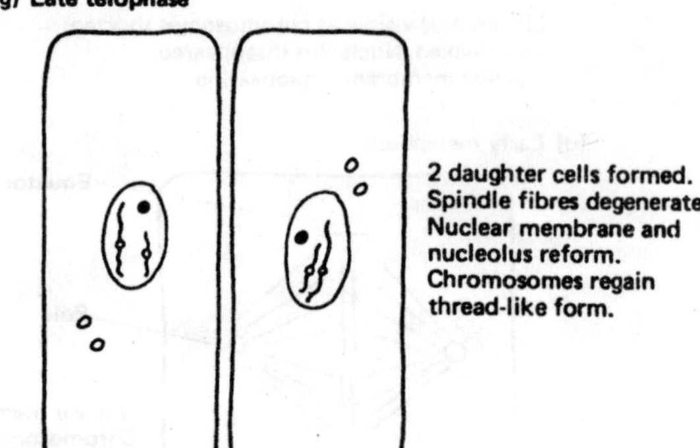

Figure 4.1: (Figure Contd.) Stages of mitosis.

are separated from each other at cell division, one going into the nucleus of one daughter cell and the second going into the other. The daughter cells are therefore identical with each other and with their parent cell in chromosome constitution.

When the cell is preparing to divide it is said to be in *interphase.* At this stage

1. the DNA replicates so that there is sufficient for two daughter cells.
2. the cell builds up its store of energy to provide sufficient for cell division.
3. the cell forms new cell organelles to supply the daughter cells.

Cell division itself is a continuous process but for ease of description four main stages are recognized: prophase, metaphase, anaphase, telophase. The following series of diagrams represents mitosis in an animal cell and shows only two different chromosomes.

Meiosis

This results in the formation of haploid daughter cells since each receives only one of each type of chromosome instead of two. The same basic stages are recognized as in mitosis but they occur twice,

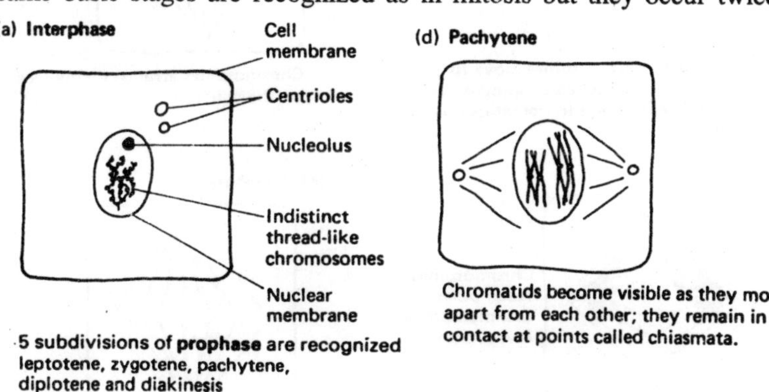

5 subdivisions of **prophase** are recognized leptotene, zygotene, pachytene, diplotene and diakinesis

Chromatids become visible as they move apart from each other; they remain in contact at points called chiasmata.

(b) Leptotene

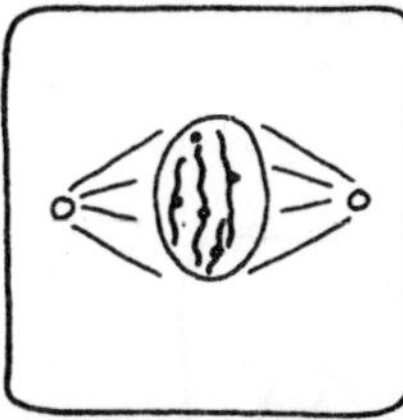

Chromosomes appear. Spindle starts to form.

(c) Zygotene

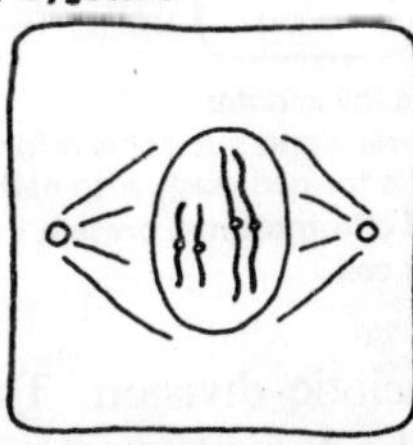

Nucleolus has disappeared. Homologous pairs of chromosomes (2 chromosomes determining the same features) associate, forming a bivalent: a process known as synapsis.

(e) Diplotene

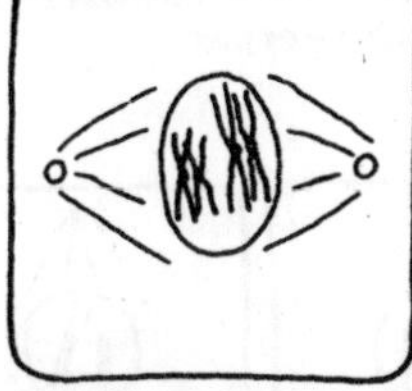

Chromatids continue to move apart as they shorten and thicken.

Diakinesis: Shortening and thickening continues; chiasmata move to ends; crossing over has occurred (see later); nuclear membrane breaks down.

(f) Metaphase I

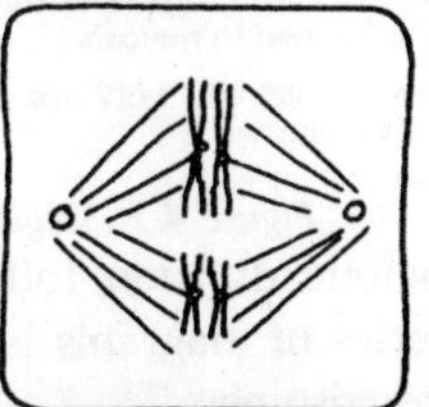

Homologous pairs of chromosomes align themselves on the equator of the spindle.

Figure 4.2: Stages of meiosis. (Figure Contd.)

(g) **Anaphase I**

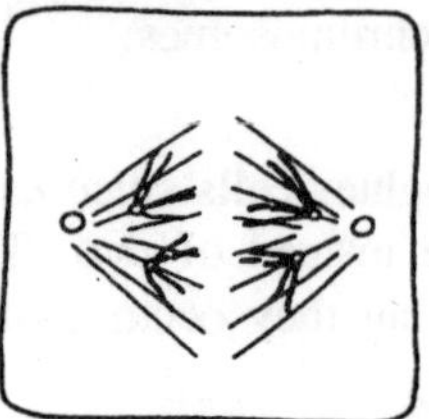

Homologous chromosomes move to opposite poles attached to spindle fibres by centromere (chromatids do *not* separate)

(h) **Telophase I**

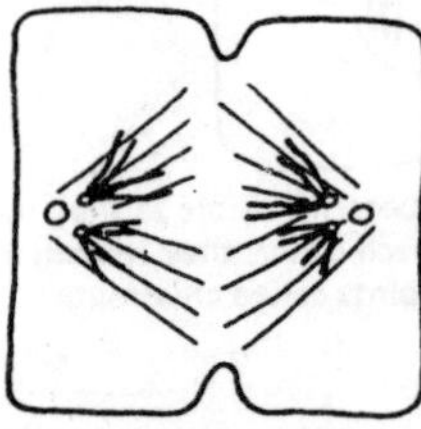

Chromosomes reach poles. Cell membrane invaginates.

There may be a short interphase or cells may move straight into the second meiotic division in which separation of the chromatids takes place.

(i) **Prophase II**

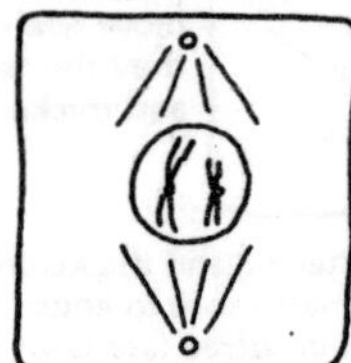

2 daughter cells.

Spindles start to form, usually at right angles to one formed in meiosis I.

In the following diagrams only one of the daughter cells is shown.

(j) **Metaphase II**

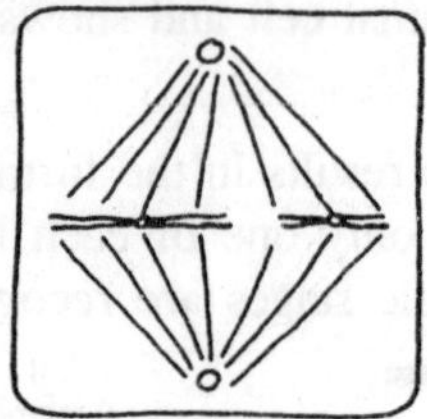

Chromosomes arrange themselves on the equator.

(k) **Anaphase II**

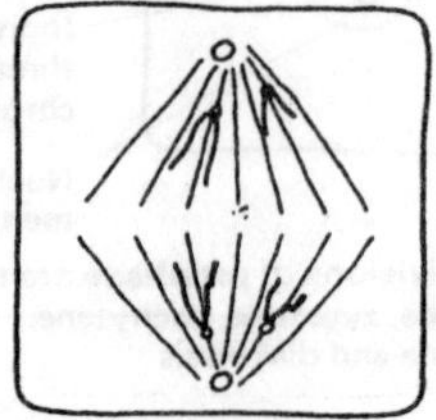

Chromatids pull apart and move to opposite poles.

(l) **Telophase II**

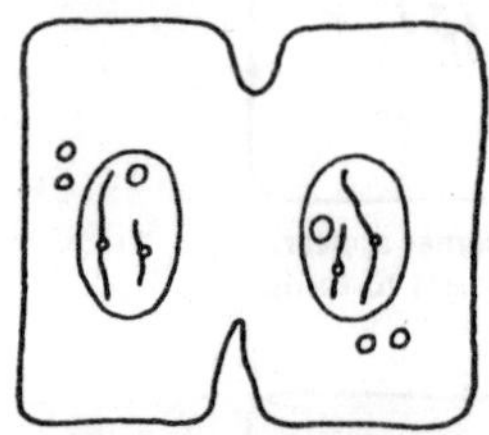

Cell membrane invaginates.
Nuclear membrane and nucleolus reform.
2 daughter cells formed, each with half the number of chromosomes present in original parent cell.

Figure 4.2: Stages of meiosis. (Figure Contd.)

i.e. first meiotic division followed by second meiotic division. The following series of diagrams is based on an animal cell containing two pairs of chromosomes.

Details of crossing over

Chiasmata may form between any two of the four chromatids and there may be up to eight chiasmata in a bivalent.

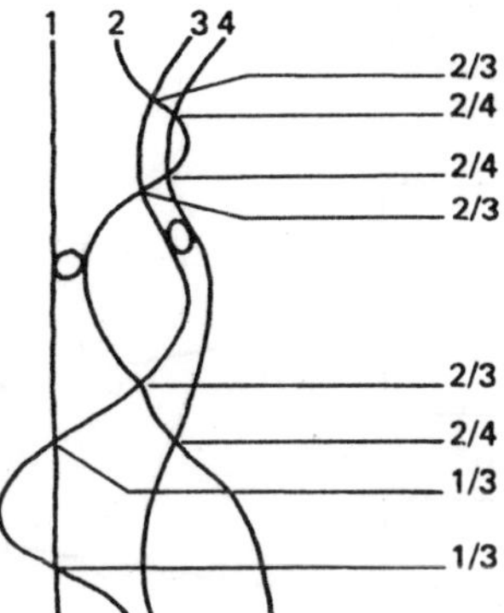

Figure 4.3: Some possible chiasmata forming in a bivalent.

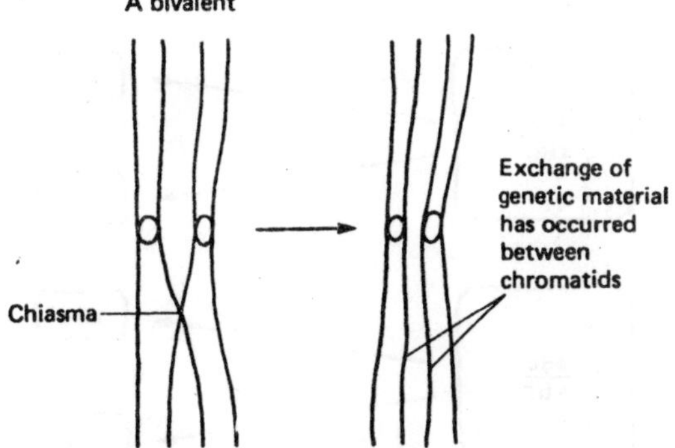

Figure 4.4: Crossover showing exchange of genetic material. The four chromatids are numbered for reference

Chiasmata have two functions:

1. to hold homologous chromosomes together while they move into position on the spindle prior to segregation.
2. crossing over (or exchange of genetic material) occurs at the chiasmata leading to increased variation, the raw material of evolution.

Significance of meiosis

1. Halving the chromosome number ensures that when gametes with the haploid number fuse to form a zygote the normal diploid number is restored.
2. Meiosis leads to increased variation:
   (a) when the haploid cells fuse at fertilization there is recombination of parental genes.

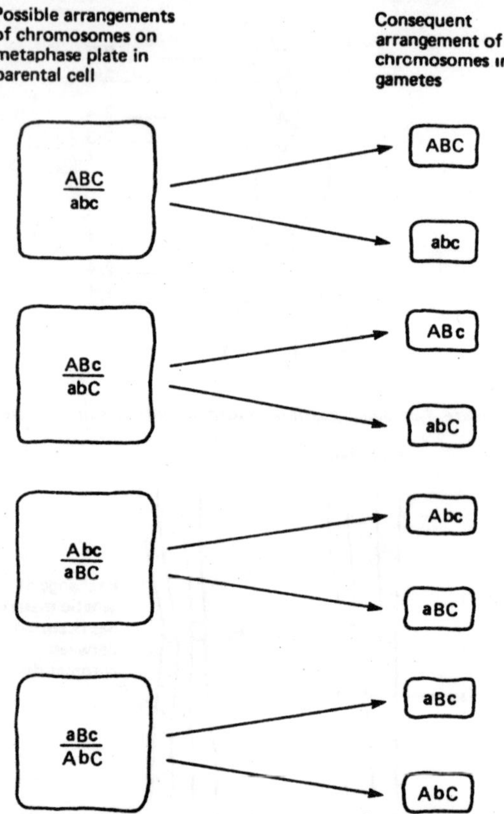

Figure 4.5: Recombination of chromosomes in gametes.

(b) during metaphase I homologous chromosomes are together at the equator of the spindle but they separate into daughter cells independently of each other.

(c) chiasmata and crossing over can separate and rearrange genes located on the same chromosome.

Table 4.1: Differences between mitosis and meiosis

Mitosis	Meiosis
One division of nucleus and one of chromosomes	Two divisions of nucleus and one of chromosomes
Chromosome number remains constant	Chromosome number halved
No association of homologous chromosomes	Homologous chromosomes associate in pairs

No chiasmata or crossing over	Chiasmata and crossing over occur
Two daughter cells formed	Four daughter cells formed (tetrad)
No variation (unless mutation occurs)	Variation due to exchange of genetic material
Chromosomes shorten and thicken	Chromosomes coil but remain longer than in mitosis
Chromosomes form a single line at the equator	Chromosomes form a double row at the equator
Chromatids move to opposite poles	Chromosomes move to opposite poles

HEREDITY AND GENETICS

Underlying Principles

Organisms of the same species vary from each other within certain broad limits. Some characteristics in a population, e.g. height and weight, change gradually from one extreme to another with every conceivable intermediate represented.

This is continuous variation and is a result of the environment acting on the organism's hereditary constitution. Other characteristics, such as eye colour and blood groups, fall into a few distinct groups with few, if any representatives of intermediates.

This is discontinuous variation and is the result of the organism's hereditary constitution alone. In sexual reproduction a new individual results from the fusion of two gametes which between them contain all the necessary information for the development of a similar, but not identical individual.

If each gamete contributes information on every characteristic of the organism it follows that at least two factors must control each characteristic in the offspring. These factors may or may not provide similar information on the character.

If they do not, then either one or other factor manifests itself in the offspring or an intermediate state between the two extremes becomes apparent. If a factor does not show itself in any one generation it nevertheless retains the capacity to do so in a later one.

Points of Perspective

Questions on this topic often involve the fruit fly *Drosophila*. The ability to recognize the different mutant types of *Drosophila* and to have carried out genetic crosses between them in the laboratory would be a distinct advantage.

As the results of genetic crosses involve analysis of numerical data,'it is an advantage to have an understanding of basic statistics, including normal distributions, standard deviations, chisquared tests and possibly binomial theory.

Some examples of, and the theory behind, selective breeding as a means of achieving better crops and animal types is useful and this could be extended into the moral implications of this and other genetic engineering in humans.

Essential Information

Organisms are recognizably similar to their parents (heredity); however they are not identical (variation).

Genotype, which is the genetical make up of organisms, often determines limits, e.g. maximum height.

Phenotype is the actual appearance of an organism. It is caused by interaction between genotype and environment.

Mendelian inheritance

The Austrian monk Mendel (1822-1884) observed clearly defined characters in the garden pea. Initially he studied only one pair of contrasting characters (monohybrid inheritance).

He isolated plants that had 'pure bred' for several generations, artificially pollinated them and observed and counted the offspring (first filial or F_1 generation). He then crossed the F_1 plants with each other to give an F_2 generation.

The characteristic apparent in the F_1 was termed *dominant* and the opposing character which disappeared in the F_1 and reappeared in the F_2 was termed *recessive*. Mendel knew nothing of meiosis, genes or chromosomes. He spoke of a pair-of factors (now called *alleles*) determining a character.

For example, the gene for height is composed of the allele for tall and the allele for short. It is usual to denote the dominant allele with a capital letter (upper case) and the recessive with a lower case letter.

Let T = tall
t = short

	Pure bred tall		Pure bred short
	TT	×	tt
gamets:	T and T		t and t
Punett square			TT

	gametes	T	T	
tt	t	Tt	Tt	F_1 generation
	t	Tt	Tt	

All the offspring have the genotype. Tt and the phenotype tall. Now cross two F_1 plants, i.e. Tt × Tt

Tt

	gametes	T	T	
tt	t	Tt	Tt	F_1 generation
	t	Tt	Tt	

TT = homozygous dominant, phenotype is tall

Tt = heterozygous, phenotype is tall

tt = homozygous recessive, phenotype is short

Therefore ratio of phenotypes is 3 tall : 1 short

This 3: 1 ratio will only be apparent if samples are large enough. From these crosses Mendel formulated

The Law of Segregation

An organism's characteristics are determined by internal factors which occur in pairs. Only one of a pair can be represented in the gametes of the organism.

Test Cross (back cross)

An organism with a dominant phenotype may be homozygous or heterozygous. In order to find the genotype of an organism it is crossed with a homozygous recessive individual of the same species. The following ratios are expected.

TT × tt

Tt

	gametes	T	T
tt	t	Tt	Tt
	t	Tt	Tt

All tall

Tt × tt

Tt

	gametes	T	T
tt	t	Tt	Tt
	t	Tt	Tt

1 tall : 1 short

Dihybrid inheritance

Inheritance of two pairs of characteristics.

Let T = tall; t = short; C = coloured; c = white

Pure bred tall plants with coloured flowers (TTCC) crossed with pure bred short plants with white flowers (ttcc).

TTCC

	gametes	TC	TC
ttcc	tc	TtCc	TtCc
	tc	TtCc	TtCc

All F_1 are heterozygous with the phenotype tall and coloured. When these are selfed to give the F_2 generation:

TtCc

	gametes	TC	tc	Tc	tC
TtCc	TC	TTCC	TtCc	TTCc	TtCC
	tc	TtCc	ttcc	Ttcc	ttCc
	Tc	TTCc	Ttcc	TTcc	TtCc
	tC	TtCC	ttCc	TtCc	ttCC

Therefore in the F_2 generation there are 9 tall, coloured
3 tall, white
3 short, coloured
1 short, white

From these results Mendel formulated

The Law of Independent Assortment

Each of a pair of contrasted characters may be combined with either of another pair.

In modern terms: each member of an allelic pair may combine randomly with either of another pair.

Genetics of sex

In humans the female sex chromosomes are XX; therefore, all female gametes contain an X chromosome (homogametic). The male sex chromosomes are XY; therefore, the gametes may contain either an X or Y chromosome (heterogametic).

In birds the female is the heterogametic sex.

In *Drosophila* the female is XX and the male XY but, unlike many species, the Y chromosome is not shorter than the X, but is a different shape.

Sex linkage

Two common examples of recessive alleles linked to the X chromsome are red/green colour blindness and haemophilia.

Let X^h represent the X chromosome carrying the allele for haemophilia.

Let X" represent the X chromosome carrying the normal allele.

Female carrying haemophilia = X^HX^h

Normal male = X^HY

X^HX^h

	gametes	X^H	X^h
X^HY	X^H	X^HX^H	X^HX^h
	Y	X^HY	X^hY

This cross gives rise to 1 carrier female (X^HX^h)
1 normal male (X^HY)
1 haemophiliac male (X^hY)
1 normal female X^HX^H

Autosomal linkage

For only 23 pairs of chromosomes to determine all the different characters in a human, it is necessary that each chromosome possess numerous alleles. All alleles on the same chromosome are said to be linked; they move together from generation to generation.

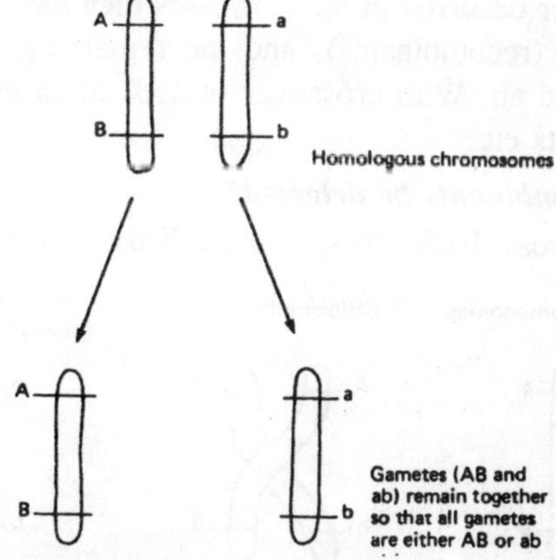

Figure 4.6: Gametes produced if A/B are linked.

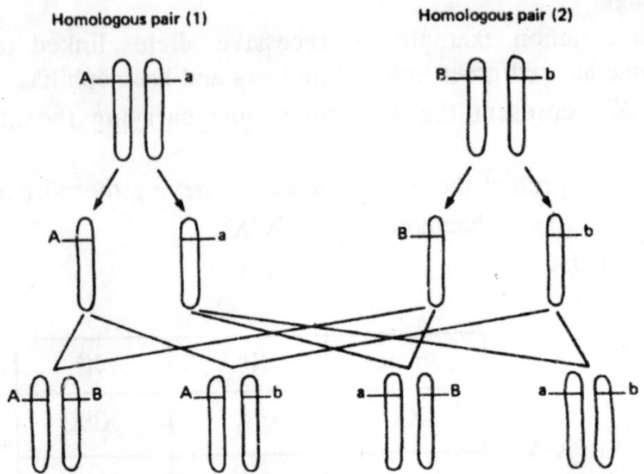

Figure 4.7: Gametes produced if A/B are on separate chromosomes.

Linkage is the association of two or more alleles so that they tend to be passed from generation to generation as an inseparable unit and fail to show independent assortment.

Crossing over and recombination

In practice the situation is not so simple because chiasmata formation and crossing over between homologous pairs during the first prophase of meiosis mean that linked alleles can be seperated.

If crossing over always occurred in the above case then all the gametes would be recombinants, i.e. Ab and aB. None would be AB and ab. This is most unlikely.

If crossover occurred in 50% of cases then half the gametes would be Ab and aB (recombinants), and the remaining half would be the original AB and ab. With crossover in 25% of cases only 25% would be recombinants etc.

How can recombinants be detected?

The test cross (back cross) is used. Take an organism of genotype

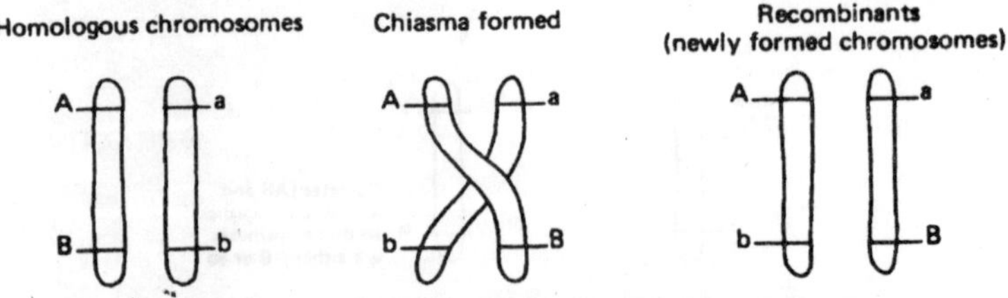

Figure 4.8: Recombination of linked alleles due to crossover.

AaBb. If back crossed to the homozygous recessive aabb the following offspring could arise:

	Gametes of heterozygous parent			
	AB	Ab	aB	ab
Gametes of homozygous parent ab	AaBb	Aabb	aaBb	aabb
Number of offspring in typical sample	125	45	40	115
	both dominant features visible	one dominant feature (A) visible	one dominant feature (B) visible	no dominant feature visible

The offspring can be determined phenotypically.

If the alleles for characters A and B were on separate homologous pairs the offspring ratio should be 1 : 1 : 1 : 1, but it clearly is not here. There are three possible explanations:

(1) A/a is not segregating in a 1 : 1 ratio as expected. To test this add up all the offspring from a gamete containing a, and this should be approximately equal to those arising from a gamete with A.

a gamete aB(40) + gamete ab(115) = 155

A gamete AB(125) + gamete Ab(45) = 170

As these figures are approximately equal, this is not the explanation.

(2) B/b are not segregating in a 1 : 1 ratio as expected. To test this follow the above procedure.

b gamete Ab(45) + gamete ab(115) = 160

B gamete AB(125) + gamete aB(40) = 165

These figures are also approximately equal; therefore this is not the explanation.

(3) Alleles are linked. To test this add the offspring from the homozygous gametes (AB and ab), i.e. (125 + 115) = 240, and compare with the offspring from the heterozygous gametes (Ab and aB), i.e. (45 + 40) = 85. These are clearly not in the ratio of 1: 1 therefore the alleles are linked and Ab and aB are the recombinants.

$$\textbf{The crossover value (COV)} = \frac{\text{Number of recombination offspring}}{\text{Total number of offspring}} \times 100$$

In the above case $\frac{85}{325} \times 100 = 26\%$ approx.

i.e. crossover occurs in about ¼ of the meiotic divisions.

It is clear that if linked alleles are at opposite ends of a chromosome any crossover will separate them and the crossover value will be high. If linked alleles are close together crossover will only rarely separate them and the crossover values will be low.

By obtaining crossover values for all linked alleles on a chromosome it is possible to determine the relative positions of the alleles and produce an accurate map of a chromosome, i.e. *chromosome mapping*.

Gene Interaction

When homozygous red and white antirrhinums are crossed the F_1 does not produce the expected result of all the dominant type. Instead all the F_1 are pink. When these are selfed the F_2 has 2 pink, 1 red and 1 white. This is called *incomplete dominance* or *allelic interaction*.

Certain breeds of poultry can be distinguished by the shape of the comb.

Rose comb — Pea comb

RRpp × rrPP

The F_1 are all walnut comb (RrPp), i.e. incomplete dominance. In the F_2 four types of comb arise:

gametes	RP	Rp	rP	rp
RP	RRPP	RRPp	RrPP	RrPp
Rp	RRPp	RRpp	RrPp	Rrpp
rP	RrPP	RrPp	rrPP	rrPp
rp	RrPp	Rrpp	rrPp	rrpp

R–P– = walnut comb

R–pp = rose comb

rrP– = pea comb

rrpp = single comb

Multiple Alleles

Sometimes more than two alleles control a particular characteristic. Only two alleles can occupy a locus on a pair of homologous chromosomes at any time. The ABO blood group system is controlled by three alleles:

allele A = production of antigen A on erythrocyte.

allele B = production of antigen B on erythrocyte.

allele O = production of no antigens.

Therefore the possible combinations are AA; AO; AB; BB; BO; OO. A and B alleles show equal dominance but both are dominant to O.

Blood group	*Possible genotype*
A	AA or AO
B	BB or BO
AB	AB
O	OO

Transmission of the alleles is in the normal Mendelian fashion.

GENETIC VARIATION AND EVOLUTION

Points of Perspective

A knowledge of the historical background to evolution and in particular the work of Charles Darwin. The evidence for evolution.

Underlying Principles

If a species is to survive it must adapt to the changing environmental conditions which occur over a period of time. As organisms disperse themselves they will encounter new conditions, to which they too must adapt.

The adaptations involve structural, physiological and behavioural changes to individuals. While interbreeding among individuals of a species continues these changes become combined in the offspring increasing variety while remaining a single species.

If however groups become genetically isolated because interbreeding fails to take place, each group may adapt to its own conditions and become sufficiently different from its neighbouring groups that interbreeding becomes impossible. Two species will then have been formed.

The mechanism by which the adaptations arise was thought by Lamarck to be the passing on to the offspring of features acquired during the lifetime of the parents.

Darwin argued that chance mutations gave rise to a population of varied individuals in which only those suited to the prevailing conditions survived and passed on their genetic features to the next generation.

Essential Information

Mutations

These are changes in genes or chromosomes responsible for much of the genetic variation which is the prerequisite for the process of evolution. They may occur as alterations in the structure of a gene or as alterations in the structure or number of chromosomes.

Most variations in gene structure are the results of errors made during the complex replication process which occurs during every cell division.

Mutation rates for a particular gene occur at a constant rate; but all genes have a low frequency rate. Frequencies vary from 1 in 10^3 to 1 in 10' replications per locus per generation.

Local environmental conditions, such as ionizing radiations or carcinogenic chemicals, increase the rates. In diploid types, the mutant gene may be dominant, recessive or intermediate in its effect. The most common mutants are recessive or partly recessive and they are deleterious (they reduce the viability of the organism and are deleted from the gene pool).

There are several ways in which chromosomes become altered, therefore giving rise to mutation. Basically, they involve a change in either structure or number.

Structural Change

This takes place when the number of chromosomes is unaltered but the linear arrangement within one or more of them is changed. Four types of structural mutation are important during evolution.

Deficiency

This results in loss of genes from one or more chromosomes.

A B C D E F G H I J — **Normal chromosome**

A B C D G H I J — **Deficiency chromosome for the genes E and F**

This gives a deficiency heterozygote and the deletion usually has deleterious consequences.

Duplication

This results in addition of genes, where a number of genes are carried twice in one of a pair of chromosomes.

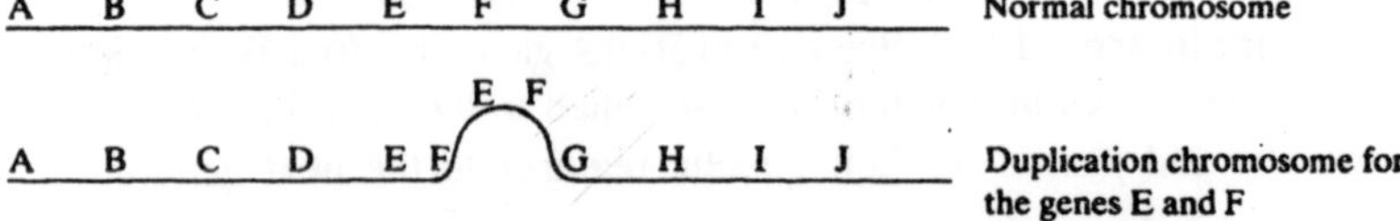

This gives a duplication heterozygote. Duplication of a gene is not usually serious but animals may show abnormalities.

Inversions

These can take place when chromosomes break in two places and the segment rejoins in an inverted position.

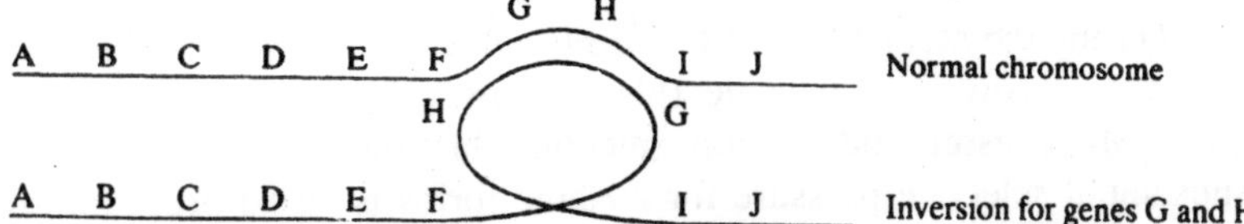

This gives an inversion heterozygote and causes a decrease in the frequency of crossing-over during meiosis.

Translocation

This is an actual change of pieces of chromosomes between two non-homologous pairs.

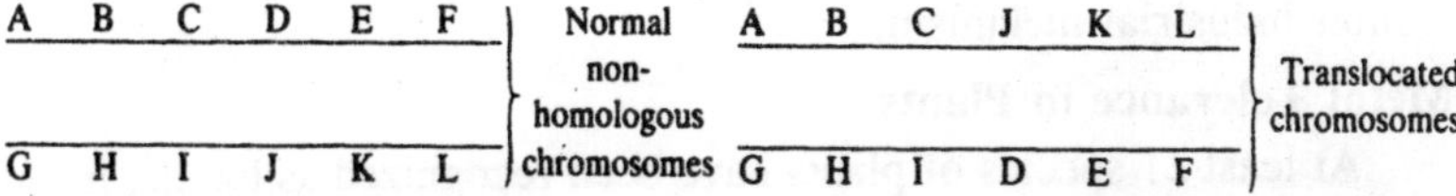

The translocation heterozygote is sterile because only those gametes with a full number of genes can give viable zygotes after fertilization.

Numerical Change

This results in a change in the normal number of chromosomes for a particular species. The most important is *polyploidy* in which the number of full sets of chromosomes is increased. Instead of somatic cells being diploid, they become tetraploid or even octoploid. A tetraploid individual will produce diploid gametes.

Triploids will produce gametes with irregular numbers of chromosomes and so will not be self-perpetuating. Polyploids are much larger than diploids of the same species and tetraploids have been very important during the evolution of plants by artificial selection.

Sometimes individual chromosomes are deleted or duplicated, giving one fewer (2n - 1) or one more (2n +1) than the normal complement. In Down's syndrome (mongolism) there are (2n + 1) chromosomes, i.e. 47 in humans.

THE RESULTS OF SELECTION PRESSURES ON MUTATIONS

Industrial Melanism

Industrial melanism in moths is an example of genetic selection induced by pollution. It was recognized in soot-polluted Manchester in 1850, when the black, melanic form of the peppered moth, *Biston betularia,* was first seen. By 1900, it had almost repla‹ ed the typical, mottled form. The increase from 1% to 99% took 50 generations (the moth has an annual life cycle).

During the second half of the nineteenth century, the rest of Britain began to receive the melanic (black) form as a result of dispersal from Manchester and natural selection in other cities. The most substantial selective pressure for melanic forms is visual predation by motheating birds, which remove the moths from trees and other vertical surfaces when they rest there during daylight.

Where surfaces are blackened by soot the melanic form is better camouflaged and predation is largely restricted to the mottled form. In areas not polluted by soot the reverse is true and the mottled form predominates. There are more than 150 species of moths known to exhibit industrial melanism.

Metal Tolerance in Plants

At least 21 species of plants have been recognized as having metal tolerant strains. Species of *Festuca* and *Agrostis* have become genetically adapted to high concentrations of zinc and lead and are tolerant on sites with a high concentration of the metals but may not survive where the concentration is low.

Where several metals are present, such as nickel and copper, plants may be tolerant to both but tolerance to one metal is genetically independent of another.

Insecticides

The widespread use of insecticides has provided a selection pressure which has resulted in adaptive changes within a very short time. It has taken only a few years, and in some cases only months, to produce strains of insects which have become completely resistant to many insecticides. At least 225 resistant species of insects have been recognized.

Many of these show resistance to DDT, dieldrin, aldrin, lindane, malathion and fenthion. The housefly *Musca domestica* has developed strains in some parts of the world which are resistant to every insecticide which can be 'safely' used, with the exception of pyrethrum compounds.

In some types of resistance enzyme induction has been responsible for the insects' ability to break down the insecticides, i.e. mutant strains 'switch on' genes which help control the synthesis of oxidoreductase enzymes to render the insecticides harmless. Many species of insects which are vectors of disease have become resistant in this way; among them the yellow fever mosquito *Aedes aegypti* and the malaria-carrying mosquitoes *Anopheles gambiae* and *A. culicifacies.*

Antibiotics

The resistance of bacteria to antibiotics has arisen by the selection of resistant mutants. As with types of insecticide resistance, some resistance to antibiotics has evolved by '*switching on*' genes to control the production of antibiotic-splitting enzymes.

All naturally occurring penicillin-resistant strains of *Staphylococcus,* as well as penicillin-resistant strains of many other species, owe their resistance to their ability to produce penicillin B lactamase.

Resistance to antibiotics has already destroyed the usefulness of several drugs. The first signs of staphylococcal resistance appeared soon after penicillin became extensively used in hospitals.

Table 4.2 Incidence of penicillin-resistant infection at a general hospital.

Date	*Total patients*	*Patients with penicillin resistant strains*
Apr-Nov 1946	99	15
Feb -June 1947	100	38
Feb-June 1948	100	59

By 1950, the majority of staphylococcal infections in all British general hospitals were penicillinresistant. Resistance has now appeared in certain staphylococci to all major antibiotics, often as a triple resistance to penicillin, tetracycline, and streptomycin.

Adaptive Immunity in Mammals

Micro-organisms and insects reproduce at such an alarming rate that it is relatively easy to visualize how they can adapt faster than man can combat them. Prolifically breeding mammals such as the brown rat and the rabbit have also evolved resistant strains capable of withstanding warfarin and myxomatosis virus, respectively.

Warfarin was developed as a rat poison. It contains di-coumarol which interferes with normal clotting mechanisms in the blood. In 1960, strains were recognized in Shropshire which were immune to warfarin. The incidence of resistant rats in parts of Wales had risen to 50% by 1970 because of the intensive selective pressure of the widespread use of warfarin.

The resistance is due to a single mutant gene and individuals homozygous for this gene are weaker than normal warfarin-sensitive rats.

Myxomatosis-resistance in rabbits became widespread in the mid 1950s. Prior to this, the virus was responsible for up to 90% mortality of rabbits in certain areas. The genetic response to this selective pressure was the production of mutants which spent a greater proportion of time above the ground, like hares.

Normal rabbits live in crowded warrens underground where the vector of the virus, the rabbit flea, can spread rapidly throughout a population. Those that spent most time above ground were favoured, whereas previously they were selected against because of predation.

Even though the rabbits were not physiologically resistant, their altered behaviour offered them protection. In the Australian population, however, a genetically resistant strain emerged after the initial epidemics. Most British rabbits are now genetically resistant to myxomatosis.

Sickle-Cell Anaemia

This is caused by a mutation which resulted in the incorporation of an incorrect amino acid at one point in the protein chains of the haemoglobin molecule. The mutated gene is known as Haemoglobin S or HbS and is recessive.

Thus only those persons having the sickling gene from both parents (i.e. homozygous for the gene) suffer acutely from the disease. The disease is an often fatal form of anaemia which is relatively common in West Africa.

Whenever the blood cells of a victim encounter a low level of oxygen, as in the venous blood of tissues, they are liable to collapse to a sickle shape and may form blockages and other complications in blood vessels. Sufferers have only a 20% chance of surviving to maturity as compared with normal people.

Heterozygous individuals suffer sickling also when the oxygen tension falls below a critical level. Given this information, one would not expect the mutation to be favoured. Indeed, such harmful mutations are frequently eliminated. This happens rapidly where it is a dominant gene, more slowly where it is a recessive, as selection acts most often on a mutation when it appears in the phenotype.

New HbS genes arise spontaneously, so they will always be present in a population, but a percentage of them should be eliminated at each generation.

However, the observations do not bear out this hypothesis. There are large areas of Africa and Asia where the gene occurs, usually in a single dose, in 15-20% of the population. There are even communities within these regions with frequencies of 40%.

A heterozygote can be recognized because the blood cells show some sickling when the person is artificially exposed to low oxygen pressure in the laboratory. Clearly, the gene is not being removed from the population. We can only conclude that in some way HbS confers an advantage over the normal Hb gene, enough to redress the losses of genes at each premature death from anaemia.

The answer lies in resistance to malaria. Children with HbS have been shown to have a 25% better chance of surviving malarial attacks than those with normal genes. Selection is not acting to remove the genes causing sickle cell anaemia.

They are removed when homozygous but this selection is outweighed by their selective advantage when heterozygous, due to better resistance on the part of heterozygotes to malaria.

EVOLUTION AND CHARLES DARWIN

Darwin became the naturalist on HMS Beagle which sailed in 1832 to South America and Australasia. During the voyage he was influenced by his knowledge of fossils and noted their similarity to present day forms.

He also noticed the differences in South American animals and plants which could be related to differences in their environment. In the Galapagos Islands he observed that while the finches there were unique they had a general resemblance to finches which existed on the mainland of South America.

He considered that originally a few finches had strayed from the mainland to these islands, and, as they bred, they produced new types with differences that allowed them to fare better in the new conditions. After the voyage he spent over twenty years developing his views on evolution.

It is problematical whether he would have published his findings had not his theory been anticipated by Alfred Wallace. Wallace sent his theory of how evolution may have come about to Darwin, who found it was in essence, the same as his own. Darwin was advised to read a joint paper on the subject to the Linnaean Society.

This he did in 1858 and the following year published his book *On the Origin of Species by Means of Natural Selection and the Preservation of Favoured Races in the Struggle for Life.*

The Darwinian Theory

Darwin's theory is concerned with three observable facts plus two deductions.

Over-Reproduction

Malthus attempted to show that all organisms tend to increase in a geometric ratio. Evidence for the enormous reproductive capacity includes:

1. The very slow-breeding elephant: if it were to bring forth six young in a life time, and if their descendants continued to breed at this rate, in 750 years there would be 19 million elephants.
2. Many fish are capable of laying millions of eggs, e.g female cod lays about 2-3 million eggs per year.
3. Flowering plants generally produce many hundreds of seeds.

Relative constancy of the numbers of species

Darwin's second fact was that despite this tendency towards a geometric increase in numbers, the numbers of a given species tended to remain constant. Hence, many plants and animals must be destroyed at some stage of their lives.

Struggle for existence

From these two facts, Darwin arrived at the first deduction. It is obvious that if the parents produce vast numbers of offspring or over many years are capable of doing so, and yet the number of a given species remains constant over many years, then there must be some sort of struggle between the organisms.

Variation among the offspring

The third observable fact was that among the offspring of any two parents (animals and plants), there are usually slight and perhaps hardly noticeable differences. Generally, no two offspring were identical.

Natural selection/survival of the fittest

This is the final deduction arising from the third fact above. If the variation confers upon the individual some greater ability to withstand the hazards and competition of the struggle for existence, then that organism will stand a better'chance of survival, and so live to breed. He anticipated inheritance by noting that *like produces like*, i.e. that the advantageous variation will be inherited, resulting in more offspring with this variation. It is important to qualify this: the inheritance of one small variation will not by itself produce a new species.

However, the production of variations in a particular direction over many generations and their inheritance will gradually lead to the evolution of a new species.

Darwin thought that it was by the production of many small continuous variations and their *cumulative inheritance* that new species arose. Modern studies of the mechanism of natural selection via genetic variation have shown that it can be considered as strong evidence for evolution. Other evidence is as follows.

A SUMMARY OF THE EVIDENCE FOR EVOLUTION

Classification

If the species present on the earth were descended from a few simpler forms, then we could expect to be able to classify them into phyla, classes, orders, families, genera and species just as is now done.

On the other hand, it is difficult to imagine how this would have been possible were the organisms not related by descent, but each specially created according to individual plans.

Homology

When the anatomy of one vertebrate is compared with that of another, in general, resemblances are more obvious than differences. Structures in individuals of two different species are said to be homologous when they show the same type of structure, the same relation to other organs of the body, and a close similarity in their embryology and development. An example is the pentadactyl limb, which is found throughout the Mammalia.

Vestigial Organs

Many organisms have obvious structures which seem to serve no useful purpose but which continue to be developed from one generation to the next. For example, in man, the caecum, appendix and coccyx. Vestigial hind limbs occur in some snakes in the form of two small bones on each side, one representing the ilium and the other the femur. Slugs have vestigial shells embedded in their dorsal surfaces. The ancestors of these types possessed these as well developed and functional structures.

Comparative Physiology

Studies of groups of animals have shown that the relationships which their structural characters indicate are the same as those which are indicated by the degree of similarity between the chemical and physical processes which go on in their bodies.

Certain diseases affect only closely related animals or plants. For example, poliomyelitis can be transmitted to primates. Among plants, wheat rust attacks many members of the Graminae. When bacteria invade organisms they are immobilized by chemicals formed in the blood called precipitins.

These are produced when any foreign protein, such as that from tissues of any other animal, is injected. Therefore, if blood or serum from a dog is injected into a rabbit, the blood of the rabbit will develop precipitins and when the blood of this rabbit is added to the blood serum of a dog, precipitation of proteins will occur.

The precipitins developed in the blood depend on the particular kind of foreign protein which is injected. Thus, in the above example, the precipitins developed in the rabbit's blood in response to the presence of the dog's serum are most effective in causing precipitation in dog's serum, and are either entirely ineffective or much less so when added to horse's serum. Precipitin tests thus provide a means of studying relationships between animals from a physiological point of view.

Embryology

The theory of recapitulation states that organisms pass through developmental stages corresponding to adult stages of organisms lower in the scale of evolution. Each individual in its development goes through a series of changes corresponding to the series of stages which classification, comparative anatomy, and comparative physiology indicate as stages which were passed through in the evolution of the species.

The appearance of gill clefts in mammalian embryos can be explained because the gills were developed in their ancestors, where they were present not only in the embryo but also in the adult. The closer the relationships of groups of animals and plants through classification, the more do their embryos resemble each other, especially in the earlier stages.

Artificial Selection

Our most useful plants have been cultivated and our most useful animals domesticated by man since prehistoric times. There are vast differences between present-day varieties and their ancestors. The varieties have evolved as a result of man selecting examples with the most desirable qualities and maintaining the qualities from one generation to the next through selective breeding.

The principle is the same as natural selection except that it is very much quicker and the features selected may not be of survival value in natural populations. It is really a practical way of demonstrating selection in the evolution of species.

Geographical Distribution

A very large amount of information is available which shows that plants and animals are by no means evenly distributed throughout the earth. Not only have certain zones, determined by latitude, elevation, temperature, etc. their own characteristic flora and fauna, but places with the same climatic conditions in different regions of the world do not always possess the same animal forms.

Elephants for example, live in India and Africa but not in South America. New Zealand and Britain have similar climatic conditions but the native fauna and flora are entirely different. The present-day distribution of flora and fauna is not completely explained by the hypothesis of 'suitability to environment'. It can be explained if two conditions are accepted:

1. that existing animals and plants are the descendants of extinct populations which were of a more generalized type, and that these dispersed from their places of origin, becoming adapted to meet new selective pressures of the environment.
2. that with changes in the topography of the land, certain groups became isolated due to the formation of barriers, e.g. sea, mountains, etc., and evolved along different paths but were prevented from interbreeding with other populations, eventually becoming so different that they could not do so even if brought together again.

Paleontology

Geologists are able to place layers of sedimentary rocks in order of their formation and they can estimate the age of each layer from the thickness of the rock above and from carbon-dating techniques. In this way, a geological time scale can be made.

Most sedimentary rocks contain fossils. Vertebrate fossils form clear evidence of evolution. Fish, which first appeared about 360 million years ago could have given rise to amphibians whose earliest representatives appeared about 280 million years ago. These could have been the ancestors to the reptiles which made their appearance around 250 million years ago.

The earliest mammals appeared as fossils in rocks about 150 million years ago, with the earliest known bird fossils present in rocks, 140 million years old. Fossil evidence is supported by specimens which appear to form a transition between two groups.

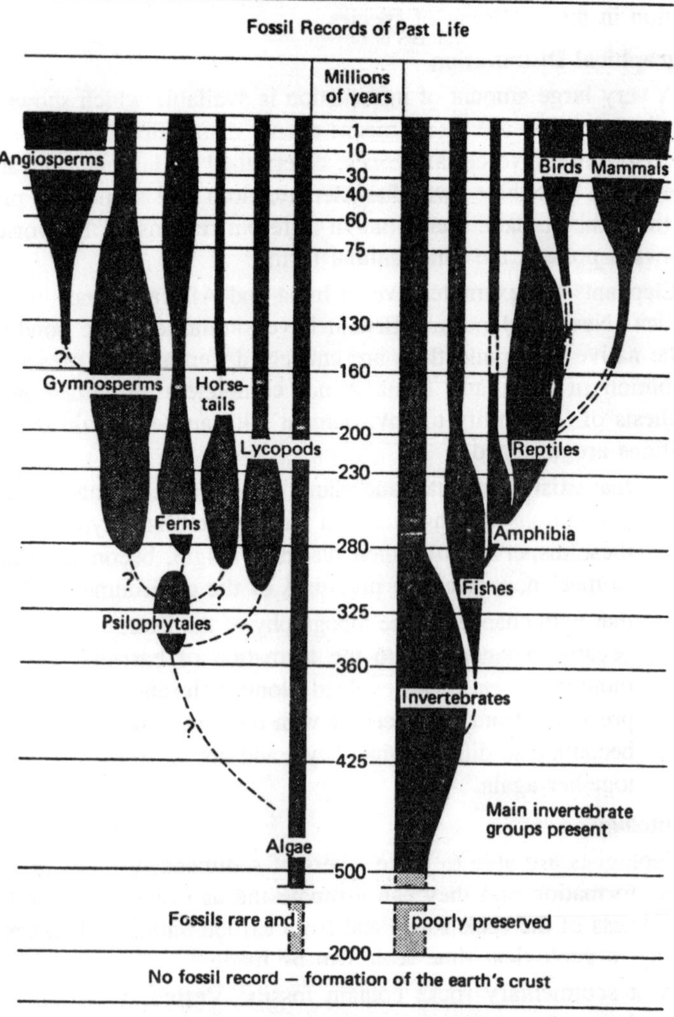

Figure 4.9: A geological time scale.

For example, *Seymouria* is a fossil which shows a combination of reptilian and amphibian features. Fossils of all the main invertebrate groups are found in rocks younger than 500 million years. The most convincing fossil evidence is found in cases where, in successive layers from one locality, a series of fossils exhibits gradual change.

Good examples of this are seen in the evolution of the oyster and in a species of sea-urchin, where complete records of most intermediate forms are seen as fossils in successive strata.

Lamarck's Theory

Lamarck published his theory of evolution in 1809 in his *Philosophic Zoologique*. He believed that the organic world is the result of the action of vital forces different from any of the physical forces operating in the organic world. His theory is developed from a series of postulates.

1. Life, by its own efforts, tends continually to increase the volume of every body which it possesses and to increase the size of its parts up to a limit which life itself determines.

   We know today that growth is the result of internal and external forces of the interaction of the organism with its environment.
2. The formation of a new organ in the body, is the result of a new need which has arisen and continues to be felt by the animal and of the new activity which this need initiates and prolongs.

   This is directly contrary to embryological evidence. For example, the vertebrate eye starts to develop in the vertebrate embryo long before there is any need for an eye, and the mammalian eye develops in the darkness of the uterus.
3. The extent of development of organs and their power of action is proportional to their use.

   It is true that use tends to increase the development of an organ and lack of use leads to degeneration.
4. All changes in or acquisition to the bodily organization of an individual during its life are transmitted to its descendants by the process of reproduction. There is no evidence of inheritance of characters acquired during the life of an individual.

The theory of Lamarck fails to give an adequate explanation of evolution because the postulates on which it is based cannot be accepted in the light of present knowledge.

5

DISPLACEMENT

LOCOMOTION AND SUPPORT IN PLANTS AND ANIMALS

Underlying Principles

Organisms initially evolved in water which gave them support. Even so skeletons were useful as a means of protection and a rigid framework for the attachment of muscles. When organisms became terrestrial it became necessary to use the skeleton for support also. For a plant the support needs to be substantial because competition for light has led to the evolution of trees over 100 m in height.

Woody plants are supported by their xylem. As xylem is a dead tissue it makes no energy demands on the organism and may therefore be massive without any disadvantage to the plant. Herbaceous plants and the herbaceous parts of woody ones rely on the hydrostatic pressure created by water entering cells by osmosis.

This pressure is termed turgor. Hydrostratic pressure is also utilized by animals. Earthworms for instance take advantage of the fact that the liquid in their coelom, like all liquids, is virtually incompressible and can act as a skeleton. Other animals use specialized tissues which, unlike those of plants, are living and require energy.

For this reason, and because most animals move from place to place, animals and their skeletons are much smaller. Where very large size has been attained the organisms have returned to water in order to gain support for their bodies, e.g. blue whale.

Locomotion is a feature of animals and some algae. The mode of nutrition in plants makes movement unnecessary because the essentials of water, carbon dioxide and light can all he obtained while remaining in one place. Indeed to move from place to place would involve being relatively small and this could put them at a disadvantage in the competition for light.

Being sessile however creates certain problems especially in the transfer of gametes during sexual reproduction. In animals locomotion is necessary in order to obtain food, although some aquatic organisms can filter food from the ambient water.

Although cilia and flagella can be used to achieve locomotion in unicellular organisms, the remainder are dependent on some musculoskeletal system. The actual arrangement of this system is dictated by the mode of locomotion, e.g. burrowing, walking, crawling, jumping, climbing, gliding, flying or swimming.

Points of Perspective

It would be an advantage to have studied and drawn all parts of the mammalian skeleton and to know the parts of different bones and their function. To have carried out microscopic examination of cartilage, bone, xylem, parenchyma, collenchyma and sclerenchyma would also be useful.

Observation of different methods of locomotion at first hand would be beneficial. Candidates with an understanding of the mechanics of support will be at a considerable advantage. It would be beneficial to have a detailed knowledge of the theories of amoeboid movement and the exact nature of muscle contraction.

Essential Information

Support in flowering plants

This is achieved in two main ways:

1 Mechanical strengthening 2 Turgor pressure

Mechanical strengthening

Some cell walls are strengthened with lignin as well as cellulose. These include collenchyma and the heavily thickened sclerenchyma. Most support is provided by the xylem sclerenchyma which constitutes the 'wood' of woody plants.

In dicotyledonous stems the mechanical tissue forms a cylinder close to the perimeter. A hollow cylinder of mechanical tissue provides the best resistance to horizontal force (e.g. wind). In roots the central core of strengthened tissue is ideally placed to resist the vertical forces met as the root pushes through the soil.

Support in animals

As an animal increases in size the soft tissues of the body require support. On the whole aquatic animals need less extensive skeletal support than terrestrial ones. There are three basic types of skeleton:

1. Exoskeleton
2. Endoskeleton
3. Hydrostatic skeleton

Exoskeleton (external skeleton), e.g. arthropod cuticle

An exoskeleton provides a more or less complete protection for the internal organs and a large area for the attachment of muscles. In arthropods the exoskelton is composed of a threelayered cuticle secreted by the underlying epidermal cells. The outer layer is a thin, waxy, waterproof layer (epicuticle).

Immediately above the epidermis is a flexible layer made of chitin which is a polymer of glucosamide and acetic acid (endocuticle). Between the two is a rigid layer of chitin impregnated with tanned proteins (exocuticle). The proteins together with lipids help to make the chitin impermeable to water.

Crustaceans also have calcium carbonate deposited in this layer to give additional strength. The inflexible plates of the exoskeleton are separated by flexible regions where the rigid exocuticle is absent; this permits movement.

Sensory hairs project through the exoskeleton which is also interrupted by the openings of various glands and the digestive,

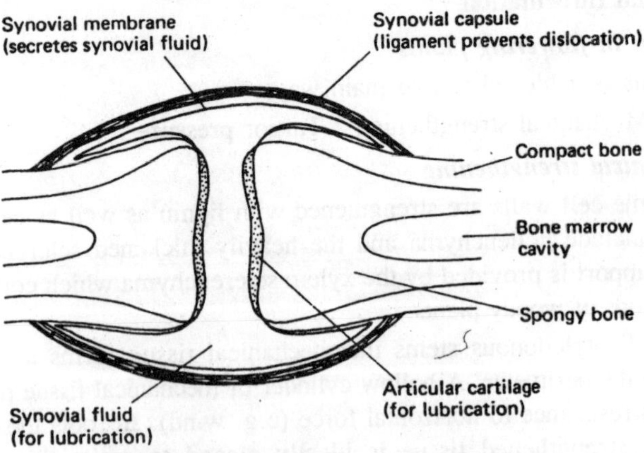

Figure 5.1: Generalized synovial joint.

respiratory and reproductive systems. The most serious limitation imposed by an exoskeleton is on growth. Chitinous plates do not grow but must be periodically shed by a process known as ecdysis (moulting).

Endoskeleton (internal skeleton)

The endoskeletons found in vertebrates are cellular although the bulk of the substance is a non-cellular matrix secreted by the cells. Cartilage forms the entire skeleton of elasmobranch fishes. It also forms the first skeleton of mammalian embryos but later most of it is converted to bone.

Cartilage combines rigidity with a degree of flexibility. Most vertebrates are supported by a bony endoskeleton, bone providing a strong, rigid, framework requiring the presence of joints if movement is to take place.

Hydrostatic skeleton

This is found in certain invertebrates, e.g. earthworm, where the coelomic fluid forms an incompressible but flexible material to aid locomotion. The muscles are arranged segmentally and act on the coelomic fluid.

Contraction of the circular muscles makes the body longer and thinner while contraction of the longitudinal muscle makes the body shorter and thicker. Chaetae on each segment anchor the appropriate part of the body so that these alternate contractions bring about movement.

Not all the circular or longitudinal muscles contract at one time; waves of contraction pass back along the body. This movement is controlled by intersegmental reflexes.

Types of Locomotion

Amoeboid movement

This is exhibited by amoeboid protozoans and cells of vertebrates such as lymphocytes. Amoeboid cells have an outer plasmalemma which adheres to the surface. Beneath the plasmalemma is a clear fluid layer which thickens into the hyaline cap at the anterior end.

The bulk of the cytoplasm is divided into a granular endoplasm surrounded by a less granular ectoplasm. For a pseudopodium to move forwards gelated proteins at the posterior end are converted to the sol state and flow forwards. They accumulate and gel near the hyaline cap.

The sol to gel theories are rather simplistic and more modern ideas employ active contraction of protein filaments to explain the changes of state. The position of the motive force differs in the three main theories:

1. The posterior end contracts and pushes the endoplasm forwards.
2. The anterior gel contracts and pulls the core of endoplasm forwards.
3. Sliding filament ratchets on the edge of the gel push the endoplasm forwards.

There is no evidence for a centre to determine the flow of the sol.

Movement by cilia and flagella

Cilia and flagella have the same basic 9 + 2 fibre arrangement. They are remarkably similar in the ciliate and flagellate protozoa and in the cells of multicellular plants (only flagella) and animals.

The outer membrane of the cilium or flagellum is continuous with the plasma membrane of the cell bearing it and the fibre system extends into the cortex of the cell, ending in a basal granule or kinetosome.

Flagella are up to fifty times the length of cilia and exhibit an effective stroke beginning at the base and travelling to the tip which exerts a pushing force on the external medium.

Cilia beat as stiff slightly curved rods in the effective stroke and have a slower recovery stroke in which they are flexible and very curved.

Groups of cilia contract in a co-ordinated way known as metachronal rhythm. It is not known how movement in cilia and flagella is brought about, but the absence of the two central fibres in cilia which only have a sensory function has suggested that these are the contractile elements. No movement appears to be possible without the basal granule therefore it is possible that this provides the source of energy required.

Muscular movements

Sliding filament theory of muscular contraction Basically this supposes that the muscle filaments do not shorten when the muscle contracts but they slide between one another. The strongest evidence for this theory comes from electron microscope studies of muscles fixed at different degrees of tension. These show:

1. The lengths of the actin and myosin filaments remain unchanged at different tensions
2. During contraction the I-band shortens

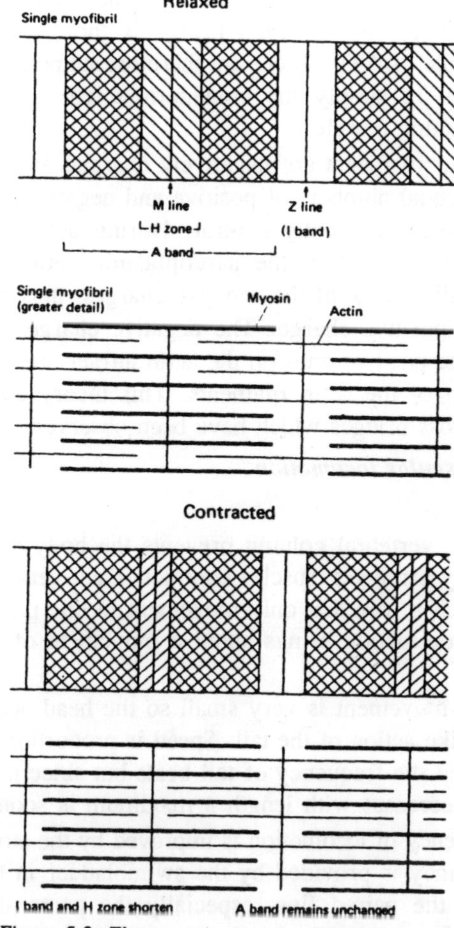

Figure 5.2: Electron micrograph of skeletal muscle.

3. During contraction the length of the A-band remains more or less constant.

There are two main views to explain the sliding filament theory:

1. There are cross bridges linking the actin and myosin filaments and sliding may be brought about by the rapid joining and breaking of these bridges in succession.

   A cross bridge joins an actin filament, contracts and then detaches; it rejoins the actin farther along the filament. The cumulative effect of this ratchet-like action is to bring about contraction. At rest the cross bridges would need to be angled in order to bring about contraction. The attachment, contraction and detachment of each bridge is thought to require one molecule of ATP. There are therefore numerous mitochondria present to supply the considerable energy demands of a contracting muscle.

2. Myosin filaments are negatively charged and actin filaments have equal numbers of positive and negative charges and are therefore electrically neutral. During activity calcium ions are released from the sarcoplasmic reticulum and these neutralize some of the negative charges, thus giving the actin a net positive charge. The negative charges on the myosin and the positive ones on the actin attract one another causing sliding of the actin filaments. This theory does not involve the cross bridges which have been observed.

Variety of muscular locomotion

Swimming

In fish the vertebral column prevents the body shortening when the segmentally arranged muscles contract. Each vertebra is joined to the next in such a way that only movement in one plane is permitted so muscular contraction brings about a movement of the body from side to side.

The head movement is very small so the head acts as a fulcrum for the lever-like action of the tail. Speed is proportional to the length of the body and the frequency of tail beats but since the frequency of the tail beat decreases with length a maximum is soon reached.

The efficiency of locomotion is improved by the streamlined shape of fish. Buoyancy is provided by the swimbladder in teleosts and by the action of the paired fins, especially the pectorals, and by the caudal fin in elasmobranchs.

Stability is provided by the various fins:

1. Yawing is prevented by vertical fins and in teleosts by the lateral flattening of the body.
2. Rolling is prevented by vertical and horizontal fins.
3. Pitching is prevented by horizontal fins.

Insect flight

The figure-of-eight wing movement of insects is brought about indirectly. There are two sets of muscles, occupying most of the volume of the thorax, whose contractions alter the shape of the thorax and thereby the wings. The very rapid wing movement of some insects cannot be explained by the thoracic muscles twitching on arrival of each nerve impulse. Instead an oscillatory contraction is set up as the walls of the thorax click in and out, like a dented tin, moving the wings as they do so. Each click deactivates the myofibrils and tension falls. The deformation of the thorax is then restored by the antagonistic flight muscles.

The most elaborate flight is possible in the two-winged flies (Diptera) which have the hind wings modified to form halteres. These act as a gyroscope to orientate the insect.

Walking and running

These forms of locomotion are particularly developed in terrestrial animals. It is important that such animals are able to balance, support the body and move.

In insects the exoskeleton supports the body and the three pairs of legs are under the centre of gravity. The head and abdomen act as counterweights. The insect moves forward by supporting its weight on a tripod formed by the anterior and posterior legs of one side and the middle one of the other while moving the other three legs forwards.

Most land vertebrates run on four legs. The weight of the body is slung from the vertebral column (often compared to a cantilever bridge). The four legs are moved in a definite pattern so that when one leg is lifted the centre of gravity lies over a triangle formed by the other three.

Therefore the order of limb movement is: left fore; right hind; right fore; left hind. This pattern can only be changed if movement is fast. Speed is determined by the length of stride x the rate of stride. Large animals have a large stride but slow rate because large muscles contract more slowly than small ones.

Limbs need to support the animal as well as propel it. Large animals may need such large legs to support their mass that their movement is slow, e.g. elephant. Fast running animals such as horses have the following adaptations to increase their speed:

1. High lever ratio of muscle to bone in leg, i.e. long leg and short muscle, attached to bone close to shoulder and hip joint.
2. As many joints as possible are moved in the same direction.
3. The bulk of the limb is reduced by:
   (a) eliminating unnecessary digits
   (b) reducing other digits
   (c) confining muscle to where movement is least.

Bipedal locomotion requires a high degree of balance and is compatible with high speed since it represents reduction of the limbs to a minimum.

6

BIOLOGICAL SOLVENT

PROPERTIES AND IMPORTANCE OF WATER

Underlying Principles

Life on earth originated in water, which remains essential to all forms of life. It is the most abundant liquid on earth and makes up more than half of every living organism, even as much as 95% of some. Probably more than any other substance it determines the distribution, anatomy and physiology of organisms.

Abundant it may be, ordinary it certainly is not. It possesses some unusual chemical and physical properties owing to the hydrogen bonds formed between each water molecule. These special properties make it an ideal constituent of, and medium for, living things. Its roles are mainly ones of lubrication, a solvent and support.

Points of Perspective

Questions exclusively about water are often concerned with its importance to living organisms and the functions it has within them. Candidates will benefit from being able to give not only a wide range of functions, but also a wide range of examples, supported by facts and figures. It would be useful to understand the physical chemistry of a water molecule including its atomic structure. Detailed knowledge of ionization, pH, buffers and hydrogen bonding would all help the overall understanding of the properties of water.

Essential Information

The water molecule (hydrogen bonding and polarity)

Naturally occurring water consists of 99.76% by weight of $^1H_2{}^{16}O$. The remainder consists of various isotopes, e.g. ^{2H} and ^{18}O. The commonest of these is ^2H (deuterium) which is most often found with the normal hydrogen atom as HDO, though occasionally as D_2O.

Both are called 'heavy water' and have a deleterious effect on living organisms. The molecule of water consists of the two hydrogen atoms bonded to the oxygen atom covalently (by the sharing of electrons).

Although the molecule overall is neutral the oxygen atom retains a slight negative charge and the hydrogen atoms a slight positive one. Such molecules are termed *polar*.

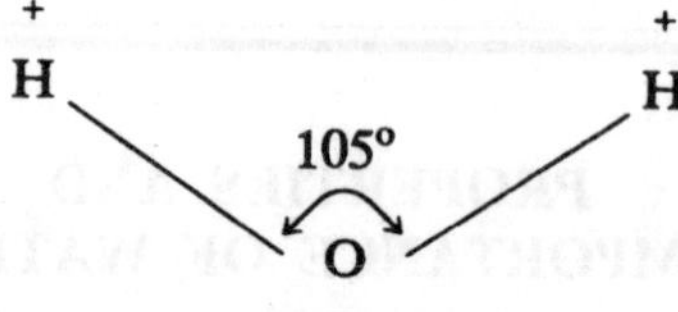

This polarity of molecules causes them to be attracted to each other by their opposite charges. These attractive forces form what are called *hydrogen bonds*. Although very weak and shortlived, such bonds collectively constitute an important force which holds water molecules together and makes water a much more stable substance than would otherwise be the case.

Properties of Water

Cohesion and surface tension

Cohesion is the tendency of molecules of one substance to hold together by mutual attraction. The hydrogen bonding of water results in strong cohesive forces. One effect of this is that the surface of a drop of water will assume the smallest possible area, and the drop therefore forms a sphere.

The water molecules at the surface are drawn in towards the body of the drop forming a skin-like layer of molecules at the surface. This force is called surface tension. Insects walking on the surface of water and the movement of water up plants are two biological processes than can occur as a result of the cohesive properties of water molecules.

Adhesion and capillarity

Adhesion is the attraction of molecules of different compounds to one another. That water clings readily to other molecules is responsible for the upward movement of water when a small bore tube is dipped in water. This phenomenon is called capillarity.

Xylem vessels of diameter 0.02 mm can, in theory, support a column of water 1.5 m by capillarity forces. One of its main biological effects is the upward movement of water in the soil.

Thermal capacity (specific heat)

Another consequence of hydrogen bonding in water is that much heat is needed to cause increased molecular movement and hence gas (steam) formation. The heat energy must first be used to break the hydrogen bonds.

For this reason the temperature of water rises only very slowly for a given amount of heat added, when compared with other substances. Similarly it cools more slowly. In all it is thermally stable and so biochemical reactions in a water medium are not subjected to large temperature fluctuations and can take place at a more constant rate.

Were it not for hydrogen bonding, water would be a gas at normal environmental temperatures and life as we know it could not exist. For the same reasons much heat is needed to evaporate water and therefore even the evaporation of a small amount of water from the surface of an organism has a large cooling effect, e.g. sweating.

Density

Water has its maximum density at 4°C. Unlike most other substances it is less dense as a solid than as a liquid. It freezes from the top downwards and the ice that forms at the surface can insulate the warmer water below from the colder temperatures above it. This prevents large bodies of water from freezing solid and has contributed to the survival of aquatic organisms.

Dissociation (ionization), pH and buffers

There is a slight tendency for water molecules to dissociate into ions according to the equation:

$2H_2O$	$\rightleftarrows$	H_3O^+	+	OH^-
water molecule		hydronium ion		hydroxide (hydroxyl) ion

It is simpler however to consider the dissociation as:

H_2O	$\rightleftarrows$	H^+	+	OH^-
water molecule		hydrogen ion		hydroxide (hydroxyl) ion

In a litre of water this dissociation produces 1/10 000 000 (10^{-7})

mole of hydrogen ions. This is equivalent to a pH of 7 which is neutral. If the concentration of hydrogen ions was greater, say 1/1000 (10^{-3}) mole hydrogen ions per litre, the pH would be 3 and the solution would be acid.

Any pH below 7 is acid, any above is basic. An acid is therefore a substance that donates hydrogen ions and a base is a hydrogen ion acceptor. Note that the pH scale is not linear but logarithmic.

A buffer solution is one that retains a constant pH despite the addition of small quantities of acids or bases. Buffers contain both hydrogen ion donors and acceptors. Bicarbonate ions may act as an acceptor.

$$\underset{\text{bicarbonate ion}}{HCO_3^-} + \underset{\text{hydrogen ion}}{H^+} \rightleftarrows \underset{\text{carbonic acid}}{H_2CO_3}$$

or a donor

$$HCO_3^- + OH^- \rightleftarrows CO_3^{2-} + H_2O$$

The removal of OH^- allows more water to dissociate and so produce more H^+

$$H_2O \rightleftarrows H^+ + OH^-$$

Bicarbonate salts and phosphate salts are responsible for the buffering of the blood, maintaining its pH at a constant 7.4. Apart from dissociating itself, water readily causes the dissociation of other substances placed in it. Thus it is an excellent solvent.

Colloids

When two substances are added to each other they may separate out or forma mixture. If the mixture consists of one substance finely dispersed throughout the other it is termed a colloid.

The substance that is dispersed is usually of a high molecular weight and does not readily diffuse through a semipermeable membrane. An example of a colloid is jelly, which comprises the protein gelatin finely dispersed in water.

Diffusion

This is the net movement of a substance from a region where it is more highly concentrated to a region where it is at a lower concentration, due to the motion of its constituent particles.

Generally a slow process, its rate is increased when it occurs over a large area, through a short distance and over a large concentration gradient.

Osmosis

Although water moves freely through cell membranes, the substances dissolved or dispersed in it may pass through more slowly, or not at all. Membranes which act in this way are termed *semi-permeable*. If two equal volumes of water are separated by a semi-permeable membrane, water molecules will move from one to the other in both directions.

The tendency of water molecules to move from one side to the other is called the diffusion pressure. The diffusion pressures are equal on each side of the semi-permeable membrane and an equilibrium is established where the movement from one side to the other is exactly counterbalanced by an equal movement in the reverse direction.

If a substance which cannot permeate the membrane is added to one side only, it effectively impedes the movement of water molecules on that side and thereby reduces its diffusion pressure. The solution is said to have a *diffusion pressure deficit* (DPD). The greater the concentration of the solution, the greater the DPD. Water moves from pure water, which has a DPD equal to 0, into the solution by the process of *osmosis*.

The pressure which must be applied to prevent this movement is called the *osmotic pressure (OP)*, i.e. it is the pressure which must be applied to prevent osmosis into a solution, when separated from pure water by a semi-permeable membrane. In practice the term can apply to solvents other than water.

Because this is a special situation and most solutions do not actually

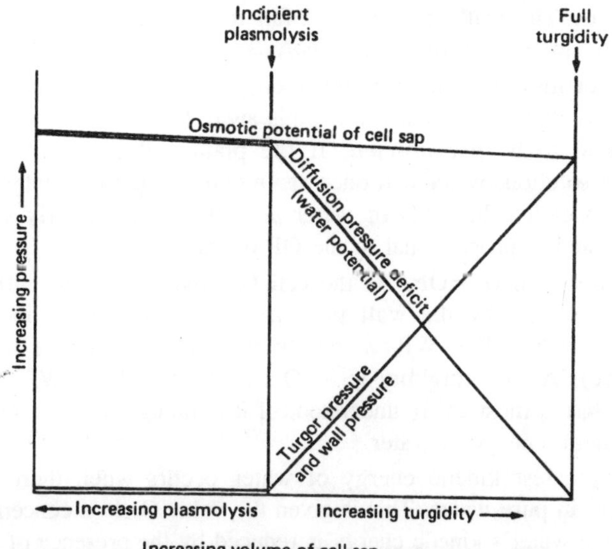

Figure 6.1: Diagram showing osmotic relationships within a plant cell.

exert a pressure, the term *osmotic potential* is more accurate. While numerically equal, osmotic pressure is a positive value, whereas osmotic potential is a negative one.

When two solutions have the same OP they are said to be *isotonic*. When one solution has a higher OP (more concentrated) than another it is said to be *hypertonic* to it. The one with the lower OP (more dilute) is said to be *hypotonic*.

Osmosis and Plant Cells

For practical purposes a plant cell can be considered as a solution of salts and sugars in the vacuole surrounded by a semi-permeable membrane (tonoplast, cytoplasm and plasmalemma) and a slightly elastic but completely permeable cell wall.

If a single plant cell were placed in pure water, the water would enter by osmosis; the vacuole would swell and push on the cell wall. This outwardly directed pressure is called the *turgor pressure*. The wall would expand a little, but would increasingly oppose the entry of water by pushing inward on the cell. Such inwardly directed pressure is called the *wall pressure (WP)*.

When the wall pressure equals the turgor pressure there will be no net entry of water. At this point the cell is fully turgid. If the same plant cell is placed in a hypertonic solution, water will leave it by osmosis. The wall pressure will fall to zero, at which point the cell is said to be at *incipient plasmolysis*.

Further loss of water from the vacuoles will cause the protoplast to shrink so that the plasmalemma pulls away from the cell wall. The cell is said to be *plasmolysed*. If the plant cell is returned to a hypotonic solution, water will once again enter. The force with which it enters is called the DPD or *water potential* (ψ) (formerly suction pressure) and is proportional to the OP of the cell sap.

Initially then ψ =OP. As the cell becomes turgid the influx of water is resisted by the wall pressure, and hence after incipient plasmolysis ψ = OP + WP (ψ and OP are normally negative and WP is positive). At full turgidity ψ = O and hence OP = WP. Water potential (ψ) is most easily understood if it is thought of as a measure of the kinetic energy of water.

The greatest kinetic energy of water occurs when there is no solute, i.e. in pure water. This is given the value O. In a concentrated solution the water's kinetic energy is reduced by the presence of solute molecules. Such a solution therefore has a lower water potential, i.e. it is more negative.

BLOOD AND CIRCULATION

Underlying Principles

As animals grew in size and complexity, tissues and organs with specific functions developed, each organ dependent on the others for some essential process or chemical. The need arose for a system to transport materials, especially food, oxygen, carbon dioxide and wastes, between the various organs.

The system must service every cell either by flowing freely over them (open system) or having capillaries within diffusing distance of them (closed system). The fluid must be circulated by cilia, body movements or a specialized pump, the heart. The heart takes many shapes and forms depending on the organism and its environment.

It may be tubular (insects), two chambered (fish), three chambered (amphibians) or four chambered (birds and mammals). Both the beating of the heart and the distribution of the blood must be carefully controlled in order to meet the varying demands placed upon it at different times.

The blood itself must be capable of carrying large volumes of many different substances, particularly oxygen. Pigments for carrying oxygen have developed and are either suspended in the plasma or, if there is a possibility of removal during excretion, in special cells.

In addition the blood contains the body's defence and immune system because it is ideally situated to fight infection in all parts of the body. Finally the liquid nature of the blood, necessary for transportation, creates problems if the body is damaged and leakage occurs. The blood has the capacity to clot in such circumstances.

Points of Perspective

It would be an advantage for candidates to have seen by dissection or demonstration the heart and major blood vessels of a fish and a mammal. It would likewise be an advantage to have dissected or have had demonstrated the heart of a mammal such as a pig or a sheep. A thorough understanding of the antigen-antibody immune system and its role in blood groups, immunization against disease and taxonomy would be beneficial. The candidate should know something of the history of blood and circulation, in particular the role of William Harvey (1578-1657) in the discovery of the circulatory system, the first vaccination by Edward Jenner in 1796 and the discovery of blood groups by Karl Landsteiner in 1900.

Essential Information

Blood makes up 7-10% of the body weight of mammals. It comprises

45% corpuscles (cells) and 55% plasma. The main components are:

1. Erythrocytes (red blood cells)
2. Leucocytes (white blood cells)
3. Platelets
4. Plasma

Erythrocytes

The main function of red blood cells is the transport of respiratory gases. Because the solubility of oxygen in water is low (0.58 cm³/100

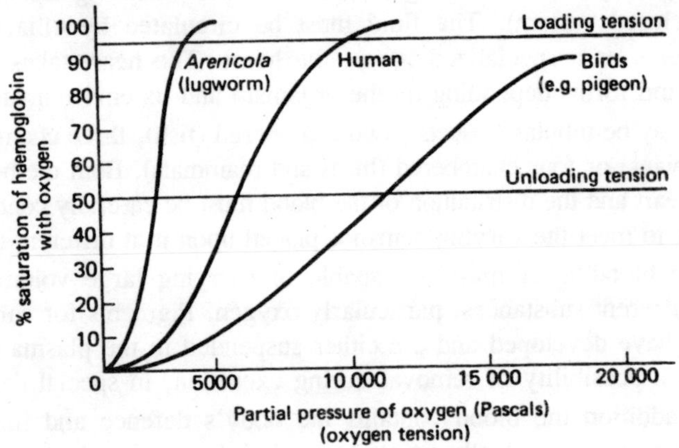

Figure 6.2 (a): Oxygen dissociation curves for haemoglobin in three organisms.

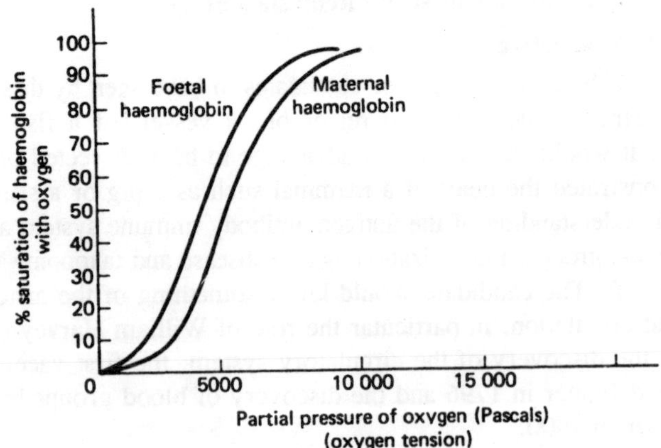

Figure 6.2 (b): Comparison of oxygen dissociationcurves for foetal and maternal haemoglobin of man.

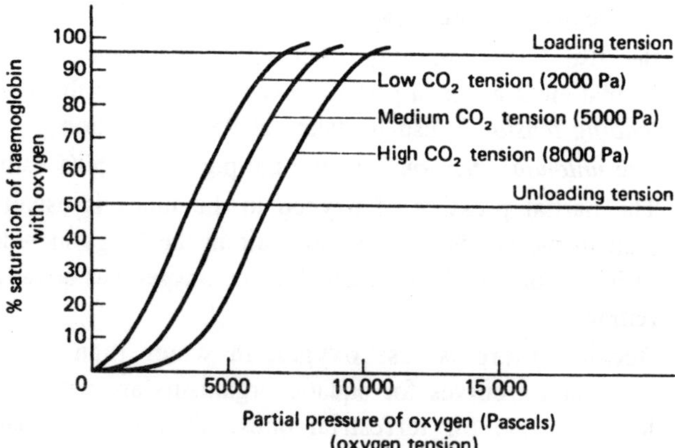

Figure 6.2 (c): Oxygen dissociation curves at different carbon dioxide concentrations—the Bohr effect.

cm³ at 25°C), *active* organisms have developed substances—respiratory pigments—with a high affinity for oxygen.

Table 6.1: Respiratory pigments.

Pigment	*Location of pigment*	*Colour*	*Group of animals*
Chlorocruorin	Plasma	Green	In some polychaetes (contains iron)
Haemocyanin	Plasma	Blue	Molluscs, some crustacea, arachnids (contains copper)
Haemoerythrin	Corpuscles	Red	Some nematodes and polychaetes (contains iron)
Haemoglobin (contains iron)	Corpuscles	Red	Vertebrates
	Plasma		Some annelids and molluscs

Respiratory pigments need to carry sufficient oxygen to supply actively respiring tissues but at the same time they must be able to release oxygen readily to those tissues.

Notes on Figure 6.2a

1. When a respiratory pigment is exposed to a gradual increase in oxygen tension it absorbs oxygen rapidly at first and then

progressively more slowly.

2. Because 100% saturation of haemoglobin does not occur at normal environmental partial pressures of oxygen the *maximum loading tension* is usually taken at 95% saturation.
3. The *unloading tension* is where the pigment is 50% saturated.
4. The partial pressure of oxygen in the lungs is usually less than in the atmosphere because air in the lungs includes the residual volume air from which some oxygen has already been removed.
5. Because there is less oxygen in water than in air the dissociation curves for aquatic organisms are usually to the left of those for terrestrial ones. This is accentuated in *Arenicola* which lives in muddy waterlogged burrows where circulation of the water is limited.
6. The dissociation curve for birds is to the right of that of man because the oxygen is given up to the respiring tissues by the pigment more readily, i.e. at a high partial pressure of oxygen. This is necessary if birds are to obtain oxygen fast enough to allow the high metabolic rate required by flight.

Notes on Figure 6.2b

The curve for foetal haemoglobin is to the left of the mother's because if the foetus is to obtain oxygen through the placenta its haemoglobin must have a greater affinity for oxygen than the mother's haemoglobin.

Notes on Figure 8.2c

1. The partial pressure of carbon dioxide in the blood alters the oxygen dissociation curve - the Bohr effect.
2. Where there is a higher concentration of carbon dioxide, i.e. in respiring tissues, the haemoglobin will release its oxygen more readily. Therefore a higher partial pressure of carbon dioxide in the blood shifts the oxygen dissociation curve to the right.
3. This is the main factor responsible for oxygen being taken up in the lungs and released to respiring tissues.
4. Animals living in regions of low oxygen tensions are less sensitive to carbon dioxide effects as any shift to the right would mean they could not absorb oxygen from the low partial pressure of oxygen in their environment.

Carbon dioxide transport

Blood takes up carbon dioxide at increasing rates as the carbon dioxide concentration increases. The carbon dioxide is carried in three ways:

Formation of bicarbonate compounds

Respiratory pigments do not carry just oxygen but also some carbon dioxide. Because haemoglobin can form a potassium salt, carbon dioxide can combine in the following way:

$$\underset{\text{carbon dioxide}}{CO_2} + \underset{\text{water}}{H_2O} \rightleftarrows \underset{\text{carbonic acid}}{H_2CO_3}$$

$$\underset{\text{carbonic acid}}{H_2CO_3} + \underset{\text{potassium salt of haemoglobin}}{KHb} \rightleftarrows \underset{\text{Potassium bicarbonate}}{KHCO_3} + \underset{\text{reduced haemoglobin}}{HHb}$$

Direct combinations with carbon dioxide

Carbon dioxide may combine with amino groups in the protein part of the haemoglobin molecule.

$$Hb—N\begin{matrix}H\\H\end{matrix} + CO_2 \rightleftarrows Hb—N\begin{matrix}H\\COO^-\end{matrix} + H^+$$

These carbamino compounds carry 2-10% of the carbon dioxide in the blood.

Carbonic anhydrase

This is an enzyme found in the corpuscles and it increases the rate of the following reaction:

$$H_2O+CO_2 \xrightleftharpoons{\text{carbonic anhydrase}} H_2CO_3 \rightleftarrows H^+ + HCO_3^-$$

The dissociation of the H_2CO_3 would lead to a rise in hydrogen ion concentration, especially since the HCO_3^- readily diffuses out of the corpuscle into the plasma.

However, the balance is restored by the movement of chloride ions from the plasma into the corpuscle. This is called the chloride shift.

Leucocytes

These are able to migrate out of the blood system and their main function is defence of the body, which they achieve in two main ways:

1. Phagocytosis: the first leucocytes to arrive at the site of an

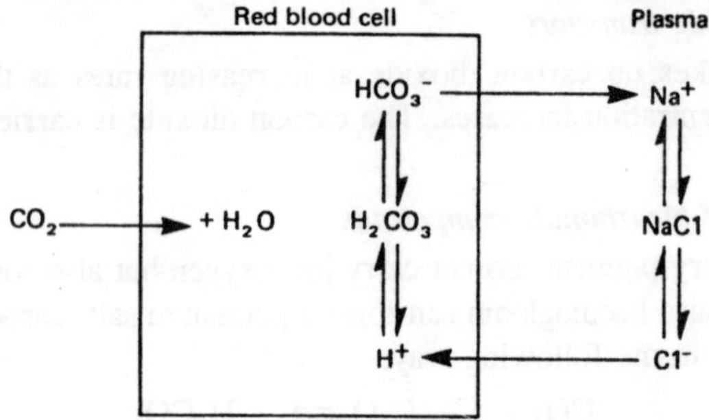

Figure 6.3: The chloride shift.

injury are usually neutrophils and they engulf foreign particles by phagocytosis.

2. Production of antibodies: an antigen is a substance which causes the body to manufacture specific proteins which agglutinate, precipitate, neutralize or dissolve foreign organisms and materials. The specific proteins so manufactured are called antibodies. Antigens are usually proteins but may be other large molecules such as polysaccharides. An antigen stimulates a lymphocyte to produce antibodies which act against foreign particles in one of a number of ways:
   (a) make them more vulnerable to phagocytic attack
   (b) interfere with the harmful activities of the foreign particles
   (c) break them down.

Platelets

The function of the platelets is in clotting and an outline of the process is given below. Prothrombin and thrombin are plasma proteins.

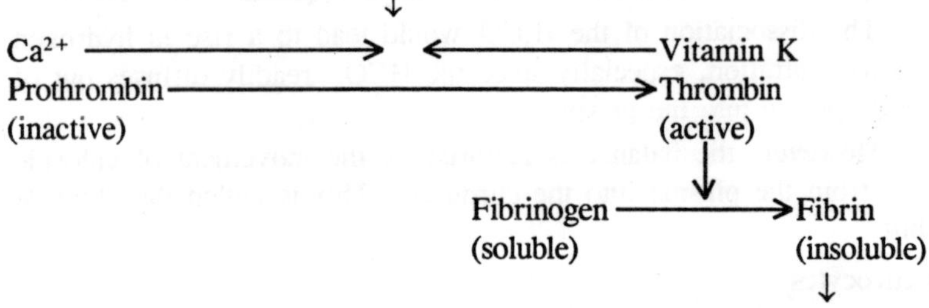

Plasma

This straw coloured liquid is an important transport medium within the body. Most of the carbon dioxide combined as HCO_3^- in the red blood cells is carried-in the plasma. It also carries:

1. the end products of digestion, especially glucose and amino acids
2. waste products, especially urea and uric acid
3. hormones
4. heat from very actively respiring tissues such as the liver and muscles to cooler parts of the body.

Blood groups

The chemical composition of the cell membrane of erythrocytes is variable. If erythrocytes of one type are introduced into the blood stream of another type they act as antigens. There are 256 known blood groups but two main grouping systems-the ABO system and the rhesus system.

ABO system

Erythrocytes may contain either of two types of antigen A or B, both, or neither. Antibodies, a and b, are present in the plasma.

Table 6.2: Blood group antigens and antibodies.

Blood group	*Antigen (in corpuscle)*	*Antibody (in plasma)*
A	A	b
B	B	a
AB	A and B	None
O	None	a and b

Incompatibility between donor's corpuscles and recipient's plasma causes agglutination and must be avoided. There seems to be little effect if the recipient's corpuscles are incompatible with the donor's plasma.

Table 6.3: British blood group frequencies.

Blood group	*% population of Britain*	*Groups from which blood can safely be transfused*
A	42	A, O
B	9	B, O
AB	3	A, B, AB, O
O	46	O

Group O is known as the universal donor and group AB as the universal recipient.

The proportions of populations with a particular blood group vary in different parts of the world, e.g. in Thailand 35% of the population are group B and in some North American Indian tribes group AB is unknown.

Rhesus system

A substance present in the blood of rhesus monkeys is also present in some humans.

Present (85% of population) = Rhesus positive (Rh^+)

Absent (15% of population) = Rhesus negative (Rh^-)

Transfusion of Rh^+ blood to an Rh^- recipient is dangerous but Rh^- to Rh^+ is safe.

During pregnancy there is some leakage of foetal blood into the mother's blood stream. If there is A, B, O incompatibility the red blood cells are quickly broken down, together with any Rh factor they may carry. If however the ABO group is compatible but a Rh^- mother is carrying a Rh^+ foetus the mother builds up Rh^+ antibodies.

This will not affect this pregnancy but the Rh^+ antibodies remain in the mother's blood stream after the birth. If there is a subsequent Rh^+ foetus in the same mother, these antibodies pass into the foetus causing agglutination and destruction of the foetal erythrocytes.

Immediately after the birth of the first child the mother may be injected with Rh antibodies which destroy the foetal erythrocytes before they cause a natural build-up of anti Rh factors in her blood.

Lymph Owing to the relatively high blood pressure in the capillaries a watery solution of low protein content, some salts, nutritive materials and phagocytic cells leaves the capillaries and bathes the tissues. This is called tissue fluid.

Having had the nutritive materials and oxygen removed and wastes added by the cells, most of the tissue fluid re-enters the capillaries by osmosis. About 1-2% is returned in separate vessels called lymphatics. This is the lymph. Lymph moves along these vessels as a result of hydrostatic pressure and respiratory and muscular movements which squeeze the lymph vessels.

The direction of flow is controlled by valves in the lymph nodes, especially in the armpits and groin. These nodes are also the site of lymphocyte production. The lymphatic vessels finally drain into the vena cava near its entrance to the heart via two major vessels, the right lymphatic duct which drains the upper right side of the body and the thoracic duct which drains the remainder of the body.

Blood Systems

These may be either open or closed.

Open blood system

In an open blood system, which occurs in arthropods and most molluscs, there are no capillaries connecting the arteries with the veins. Arterial blood passes into sinuses so that major tissues are bathed in circulating fluid.

These fluids slowly work their way back to the open ends of veins or to the ostia of the heart. The organs lie directly in the blood-filled haemocoel.

Closed blood system

In a closed blood system, as found in vertebrates and annelids, a continuous network of minute capillaries unites the smaller arteries with the veins. Nutrients are transferred to cells by fluids filtering through the walls of the capillaries into the tissue spaces.

Single circulatory system (fish)

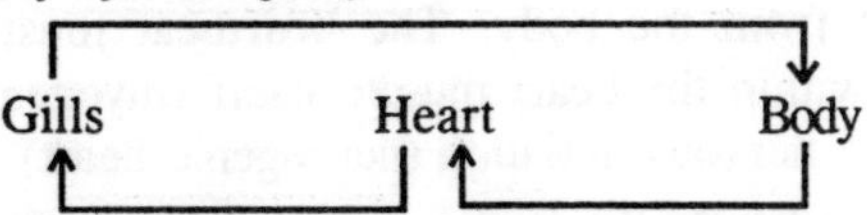

Blood passes through the heart only once per circuit of the body. Being forced through the gill capillaries lowers the blood pressure considerably and it still has to pass across another capillary system before returning to the heart.

Double circulatory system (birds and mammals)

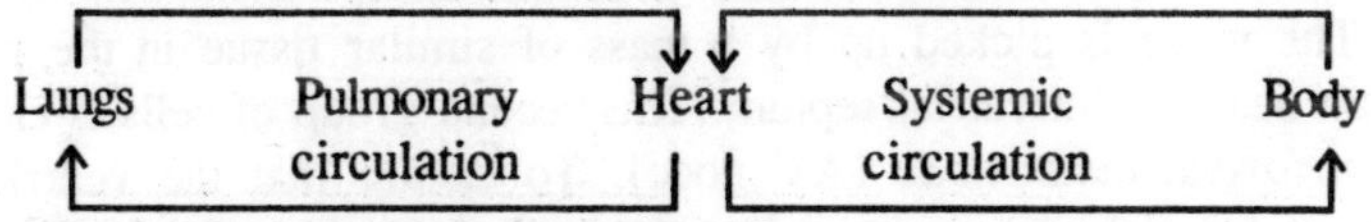

Blood passes through the heart twice per circuit of the body. This enables blood pressure to be maintained.

The Heart

The simplest form of heart is a muscular contracting region of a blood vessel, e.g. in an earthworm. In fish this contractile region is S-shaped and divided into a receiving region, the atrium, and a pumping region, the ventricle.

In amphibians there are two atria and a single ventricle, whereas in birds and mammals there are two atria and two ventricles arranged so that the left atrium and ventricle carry oxygenated blood and are completely separated from the right atrium and ventricle which carry deoxygenated blood.

The left atrium and ventricle are separated by the bicuspid or mitral valve and the right atrium and ventricle by the tricuspid valve. The aorta and pulmonary arteries have semilunar valves to prevent backflow of blood into the ventricles.

The left ventricle pumps oxygenated blood to the whole body except the lungs and has a thicker muscular wall than the right ventricle which pumps deoxygenated blood to the lungs only. The heartbeat consists of a contraction phase (systole) and a relaxation phase (diastole).

In a resting adult human the heart pumps about 70 times per minute, pumping about 60 cm^3 of blood per beat from each ventricle. The total volume of blood pumped each minute is called the *cardiac output* and varies according to the amount of physical exercise undertaken and the emotional state of the individual.

Control of Heartbeat (cardiac rhythm)

A vertebrate heart continues to beat in a co-ordinated way even when removed from the body. The heartbeat must, therefore, be initiated from within the heart muscle itself (myogenic heart) rather than by separate nervous initiation (neurogenic heart).

In myogenic hearts the beat is initiated by a '*pacemaker*' called the *sinoatrial node* (SA node)-a group of specialized cardiac muscle cells, 2 cm by 2 mm, in the right atrium, near the point where the venae cavae enter. The wave of excitation from the SA node spreads outwards causing the atria to contract, emptying their blood into the ventricles.

The wave is picked up by a mass of similar tissue in the right atrium near the interatrial septum. This second group of cells is called the *atrioventricular node* (AV node). To ensure that the ventricles contract from the apex upwards and so allow blood to be forced out through the arteries, the wave of excitation is conducted to the apex of the ventricle by the *Purkinje fibres* (bundle of His).

Factors Modifying Heartbeat (cardiac rhythm)

Chemical control

To maintain its rhythm for any length of time the heart requires the correct balance between the ions of calcium, sodium and potassium. The chemicals adrenalin and noradrenalin and drugs such as digitalis increase cardiac output. An increase in the level of carbon dioxide in blood (effectively a fall in pH) also causes an increase in cardiac output. Acetylcholine slows myogenic hearts while accelerating neurogenic ones.

Nervous control

Receptors in the aorta and carotid arteries respond to an increase in blood pressure by sending nervous impulses to the cardio-regulatory centre in the brain, which in turn sends impulses along the efferent

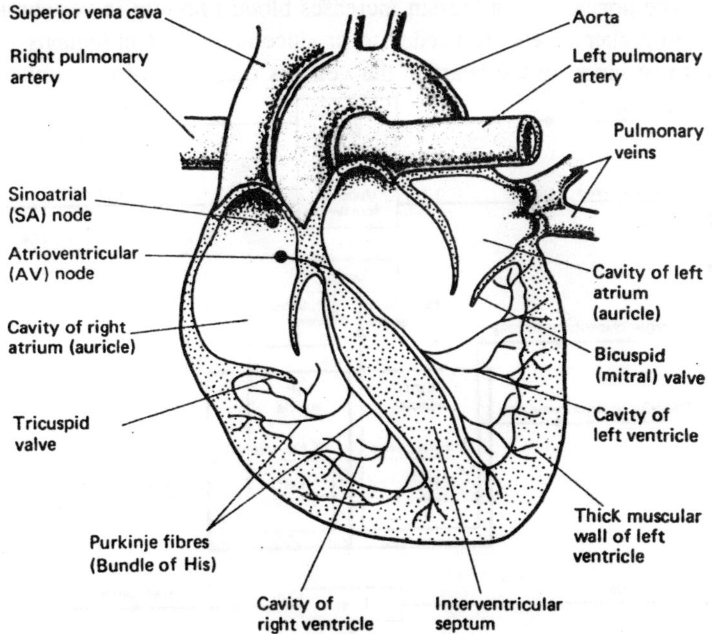

Figure 6.4: Diagram of a vertical section through the human heart (ventral view).

vagus nerve to decrease cardiac output. An increase of pressure in the right atrium causes receptors to send impulses along the afferent vagus nerve to the cardio-regulatory centre which then sends impulses to increase cardiac output. The heartbeat is therefore speeded by the sympathetic nervous system and slowed by the parasympathetic nervous system.

The maintenance and control of blood pressure and circulation

In a healthy human at rest the arterial systolic blood pressure is 120 mm Hg and the diastolic pressure is 80 mm Hg. The pressure is created initially by the contraction of the ventricles of the heart. As blood is forced into the arteries, the elastic walls expand causing distension.

The recoil of the elastic walls pushes blood away from the heart, creating distension of the artery at a point further away from the heart. (The blood is prevented from returning to the heart by the semilunar valves.) The 'pulses' continue throughout the arterial system and help to maintain blood pressure.

The hormone vasopressin increases blood pressure in vertebrates, though higher levels are needed to produce this effect in humans. The return of blood to the heart in the veins is maintained in a number of

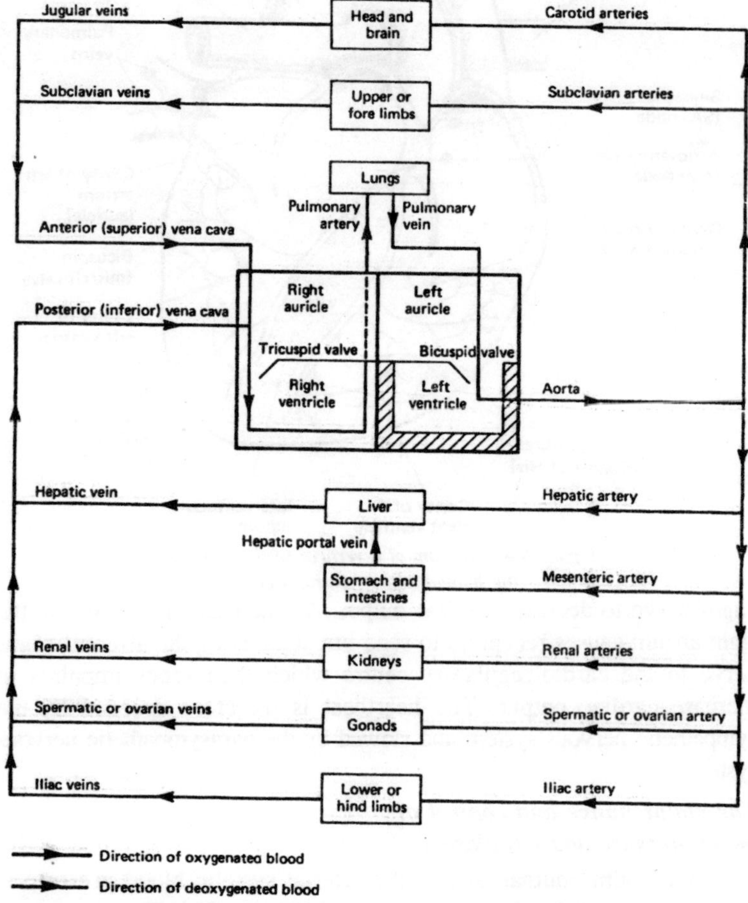

Figure 6.5: Outline diagram of mammalian circulation.

ways:

1. The residual heart pressure—usually 10 mm Hg or less.
2. The contraction of muscles squeezes veins and forces blood towards the heart; flow in the reverse direction is prevented by pocket valves.
3. Inspiratory movements -when breathing in, the low thoracic pressure helps to draw blood along the major veins towards the heart.
4. Gravity will help return blood from those regions above the heart.

The 5 l of blood in a human is inadequate to supply the maximum needs of all regions of the body at the same time. Different organs make different demands on the blood both in quality and quantity.

Table 6.4: Blood flow to various organs.

Organ	*Blood flow in em³ to the organ (when body is at rest) per 100 g of organ*
Heart	80
Liver	90
Brain	55
Muscle	3
Skin	10
Kidney	400
Remainder	2.5

Clearly, during exercise muscles require considerably more blood than they do when at rest. Control of blood flow is achieved by contraction and relaxation of muscles in the smaller artery (arteriole) walls.

Control of these muscles is by the autonomic nervous system, the sympathetic system contracting the vessels and decreasing blood flow, the parasympathetic system dilating them and increasing the flow. Apart from changes in blood flow to meet physiological needs, certain psychological events control flow in humans, e.g. fear, embarrassment and erection of the penis or clitoris due to erotic stimulation.

UPTAKE, TRANSPORT AND LOSS IN PLANTS

Underlying Principles

Most organisms comprise a variety of different cells grouped into tissues and organs. These tissues and organs have become specialized to perform particular functions. The product of one tissue may be required by another and consequently a transport system between the two is required. In addition there are certain advantages for plants in being large, for instance, they can compete more readily for light.

Some have become extremely tall (over 100 m) and have thereby obtained competitive advantage. This has meant that the organs collecting the water, the roots, are some considerable distance from the leaves that require it for photosynthesis.

Again this necessitates a transport system between the two structures. The sugars manufactured in the leaves must be transported in the opposite direction to sustain respiration in the roots. Unlike most animals, plants do not possess contractile cells such as muscle cells.

They are therefore dependent to a large degree on passive rather than active mechanisms for transporting material. Evaporation of water from leaves creates an osmotic gradient along the leaf mesophyll cells that draws water in from the xylem. As water is drawn from the xylem, the cohesive properties cause it to be pulled up in a continuous column.

An osmotic gradient in the roots brings water from the soil to the xylem. Only its entry into the xylem involves the expenditure of energy produced by the plant itself. The downward transport of sugars is less clearly understood. Certain theories, e.g. mass flow theory, favour a totally passive process, while others, e.g. transcellular strand theory,

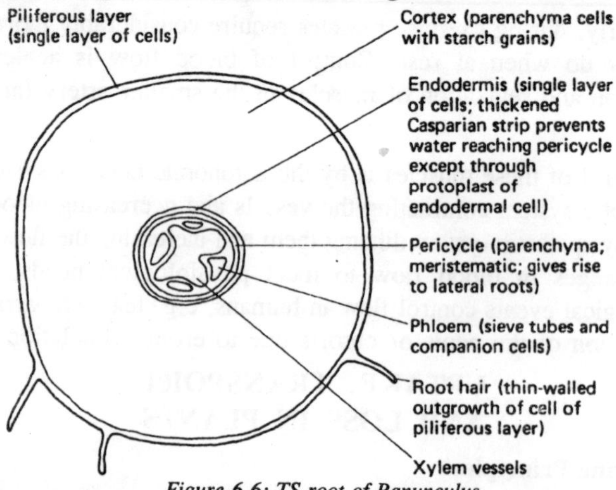

Figure 6.6: TS root of Ranunculus.

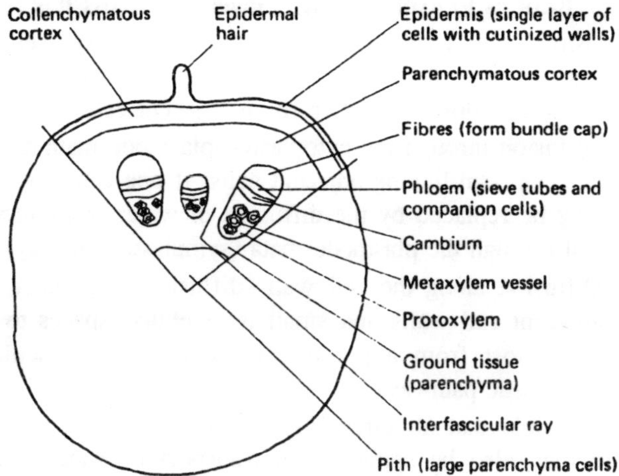

Figure 6.7: TS stem of Helianthus.

involve energy expenditure.

Points of Perspective

Candidates with practical experience of using a potometer will be at an advantage. It would be beneficial to have read widely on the theories of transport in phloem and to be able to evaluate each one critically.

A broad knowledge of xeromorphic adaptations, described with reference to specific examples would be useful. Knowledge of the problems of hydrophytes and halophytes and the methods they use to overcome them might gain one or two additional marks.

Essential Information

Water uptake by roots

Available soil water is absorbed from the water spaces between the soil particles and the water film surrounding them. Absorption is primarily by the root hairs arising from the piliferous layer of the root-a thin layer with hairs increasing the surface area. Near the apex of the root there are no hairs but the absence of a cuticle makes direct absorption possible.

The cells of the piliferous layer and root hairs contain cell sap with a more negative water potential (SP) than the available soil water thereby allowing water to move into them by osmosis.

Movement across the cortex

The cortex comprises closely packed parenchyma cells. There are three main theories to explain water movement across this region:

1. Osmosis—as water is drawn from a parenchyma cell into the vascular stele the water potential (lip) of that cell decreases and water flows into it from an adjacent cell by osmosis.
2. Diffusion through the cytoplasm—plasmodesmata are strands of cytoplasm linking adjacent cells; if one cell loses water it may be replaced by the diffusion of water from an adjacent cell through the plasmodesmata (symplastic pathway).
3. Diffusion along the cell wall—diffusion may occur through adjacent cell walls and small intercellular spaces to replace water lost from a parenchyma cell to the vascular stele (apoplastic pathway).

It is possible that all three mechanisms operate to some extent and there may also be some active transport of water within the parenchymatous cortex.

Movement into the Xylem

The vascular cylinder is separated from the cortex by a highly specialized cylinder of cells, the endodermis. Each endodermal cell, as well as having a cellulose cell wall, has a Casparian strip of suberin which prevents water passing across its radial and horizontal walls. This means that all water passing from the cortex to the xylem must pass through the cytoplasm of an endodermal cell.

As the cell ages heavy thickening covers the Casparian strip but some cells remain in the primary condition and are known as passage cells. When a stem is cut just above soil level the cut end exudes water for some time. This is evidence for water being 'pushed' into the stem from the roots by root pressure.

Such movement of water from the cortex into the xylem is thought to be an active (energy requiring) process since the application of a metabolic poison prevents it. The energy for the process may come from the oxidation of starch in the endodermis.

Movement within the xylem

The structure of xylem vessels and tracheids is given in other Section of this chapter ('Plant and animal tissues'). Xylem vessels and tracheids lack cell contents and so their hollow lumina permit an unimpeded flow of water through the plant.

Lateral movement within the xylem is possible through the pitted sidewalls of both vessels and tracheids. The forces of adhesion and cohesion maintain a long, unbroken column of water within the xylem.

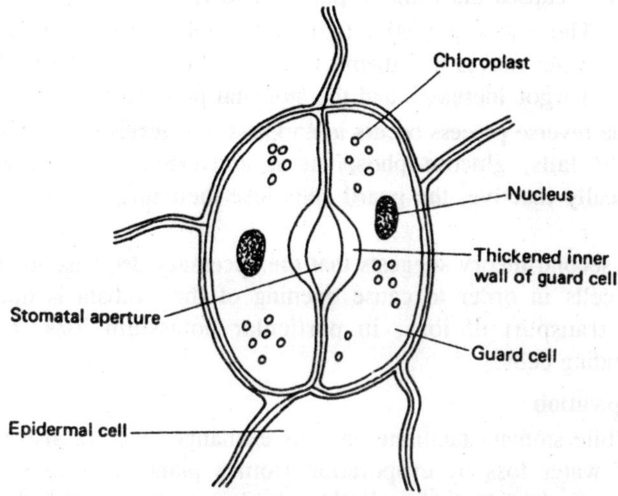

Figure 6.8: Surface view of a stoma

Movement into the leaf

The theories for the movement of water into the leaf are the same as those for movement within the cortex, i. e. osmosis and diffusion through the cytoplasm or along the cell wall. In all cases the necessary gradient is produced as water is lost from the leaf of the plant through the stomata.

Stomata

These are found in leaves and herbaceous stems. Each stoma comprises a pair of specialized epidermal cells called guard cells and an elliptical pore, the stomatal aperture.

The frequency of stomata varies with environment and species. They are generally more numerous on the abaxial (under) surface of leaves than on the adaxial (upper) side. Stomata usually open in the light and close in the dark. Although it is not without criticism the main theory to explain this is as follows:

1. In the light carbon dioxide is used in photosynthesis and so the concentration of carbon dioxide in the tissues falls. This causes an increase in the pH.
2. The rise in pH in the guard cells causes:

$$\text{starch} + H_3PO_4 \xrightarrow{\text{starch phosphorylase}} \text{Glucose-1-phosphate}$$

3. The glucose- 1-phosphate is hydrolysed

$$\text{Glucose-1-phosphate} + H_2O \rightarrow \text{glucose} + H_3PO_4$$

which causes the osmotic potential to rise.

4. The water potential (lii) of the guard cells decreases and water moves into them by osmosis from the surrounding cells. Turgor increases and the stomatal pore opens.

The reverse process occurs in darkness, i.e. levels of carbon dioxide rise, pH falls, glucose-lphosphate is converted to starch which is osmotically inactive, the guard cells lose their turgidity and the pore closes.

A second theory suggests that the necessary decrease in W of the guard cells in order to cause opening of the stomata is due to the active transport of ions, in particular potassium ions, from the surrounding cells.

Transpiration

While stomata facilitate gaseous exchange they are also the main site of water loss by evaporation from a plant. This loss of water from plants is called transpiration. It also occurs through the cuticle and lenticels. The upward movement of water from the roots is partly dependent on the pull exerted by transpiration, the transpiration pull.

Factors affecting the rate of transpiration

The rate of transpiration is dependent on the size of stomatal aperture and the diffusion gradient between the leaf and the atmosphere. The latter is affected by:

1. Humidity: the lower the humidity outside the plant the steeper the diffusion gradient between the leaf and the atmosphere. Transpiration is faster the lower the relative humidity of the atmosphere.
2. Wind speed: an increase in wind speed normally increases the rate of transpiration since saturated air is blown away from the stomatal pore and a diffusion gradient is maintained. This is partly counteracted by the cooling effect of wind.
3. Temperature: if the temperature of the air and leaf increase simultaneously, the consequent increase in vapour pressure of the leaf increases the diffusion gradient and transpiration increases.

Xerophytes

These are plants adapted to survive conditions of unfavourable water balance and they may show the adaptations shown in Table elsewhere in this chpater.

Table 6.6: Adaptations in xerophytes.

Mechanism	*Adaptation*	*Examples*	*Notes*
Reduces transpiration	Thick cuticle	Most evergreens	Reduces cuticular transpiration; often shiny so reflects sun causing leaf temperature and therefore transpiration to fall
	Depression of stomata	*Ilex* (holly), *Pinus*	Lengthens diffusion path and therefore reduces diffusion pressure gradient; may trap still, moist air
	Rolled leaves	*Ammophila* (marram grass), *Calluna*	Folding lengthens diffusion path and traps moist air
	Protective hairs	*Ammophila*	Often accompanies rolling of leaf; traps moist air
	Leaves small or absent	*Pinus*	Small and circular in cross section to give low surface area/ volume ratio and structural support to prevent wilting
		Opuntia	*No* leaves, flattened stem photosynthesizes
	Variations in leaf positions	*Lactuca sp.* (European compass plant)	Leaf positions adjusted so that sun strikes them obliquely; lowers transpiration because lowers temperature
	High osmotic potential of cell sap	Many xerophytes	Thought to reduce evaporation from cell walls
Succulence	Succulent stem and possibly leaves	Cacti	Stores water
	Diurnal closing	Many xerophytes	Reduces transpiration; requires metabolic modifications

(Table 6.6 Contd.)

(Table 6.6 contd.)

Mechanism	*Adaptation*	*Examples*	*Notes*
	of stomata		
	Shallow, wide root system	Many xerophytes	Gains maximum benefit from light rain
	Vegetative propagation well developed	Many xerophytes	Seed germination requires water
Extensive root systems		Most xerophytes	Especially developed in the most arid conditions
Resistance to desiccation	Increased lignification	*Ruscus*	Flattened stem (cladode) takes over photosynthesis
	(correspondingly	*Acacia*	Lamina lost; petiole flattened to form phyllode reduced leaves)
	allows resistance	Cacti	Leaves reduced to spines to wilting
	Reduced cell size	Many xerophytes	Less likely to wilt than fewer large ones

Table 6.7: Mineral elements necessary for plant growth.

Macro/ micro nutrient	*Mineral*	*Form in which element is absorbed*	*Function*	*Effect of deficiency*
Macro-nutrient	Nitrogen	NO_3^-	Component of amino acids, proteins, nucleotides, chlorophyll, some plant hormones	Chlorosis (yellowing of leaves); stunted growth
	Phosphorus	H_2PO_4 ATP,	Component of proteins; needed for nucleic acids and phosphorylation of sugar	Stunted growth; dull, dark green leaves
	Potassium	K^+	Component of enzymes and amino acids; needed for protein synthesis and in cell membranes	Yellow-edged leaves; premature death
	Calcium	Ca^{2+}	Calcium pectate in mid-lamella of cell walls; aids translocation of carbohydrates and amino acids; affects permeability of cell membranes	Death of growing points, therefore stunted roots and shoots
	Magnesium	Mg^{2+}	Constituent of chlorophyll; activator of some enzymes	Chlorosis
	Sulphur	So_4^{2-}	Constituent of some proteins	Chlorosis; poor root development

(Table 6.7 Contd.)

(Table 6.7 Contd.)

Macro/ micro nutrient	*Mineral*	*Form in which element is absorbed*	*Function*	*Effect of deficiency*
	Iron	Fe^{2+}	Constituent of cytochromes; needed	Chlorosis for synthesis of chlorophyll
Micro-nutrient	Boron	BO_3^{3+} or B_4O^{2+}	Aids germination of pollen grains and uptake of Ca^{2+} by roots	Diseases such as heart rot of celery, internal cork of apples
	Zinc	Zn^{2+}	Activator of some enzymes; essential for leaf formation; needed for synthesis of IAA	Malformation of leaves
	Copper	Cu^{2+}	Constituent of some enzyme systems	Growth abnormalities, e.g. dieback of shoots
	Molybdenum	Mo^{4+} or Mo^{5+}	Affects nitrogen reduction through enzyme system	Reduced yield of crop
	Chlorine	Cl^-	In osmosis and important in anion-cation balance of cells	Difficult to demonstrate
	Manganese	Mn^{2+}	Enzyme activator	Type of chlorosis; leaves mottled with grey patches

Uptake and transport of mineral salts

Like water, mineral salts are absorbed through the root hairs and other young parts of the root. The application of metabolic poisons prevents the uptake which indicates that the process is active. There is also evidence that salts are taken up selectively (proportions of minerals inside and outside the plant are different) and against the concentration gradient.

The mechanism and route of movement across the cortex to the xylem are not certain. There are a number of theories relating to the mechanism of transport across membranes but most postulate the existence of a carrier molecule, possibly a protein.

Once inside the xylem the ions travel upwards in the transpiration stream. Although xylem is principally responsible, mineral transport also occurs in the phloem and lateral movement from xylem to phloem is possible.

When the mineral ions reach the leaves and meristematic areas of the plant they diffuse out of the xylem vessels and are absorbed by the cells for various metabolic functions, e.g. building up proteins and amino acids.

Translocation of Organic Materials

Evidence for movement in the phloem

1. There are diurnal variations of sucrose concentration in the leaf and, after a time lag, these are reflected in the phloem sieve tubes.
2. If a stem is *'ringed'* so that *phloem* is removed, sugars accumulate above the ring. Dyes however move normally.
3. If a plant is given $^{14}CO_2$ and ringed to remove the phloem, the sucrose accumulating above the cut contains ^{14}C.
4. Aphids use needle-like mouthparts to obtain sugars from phloem sieve tubes. Removal of the aphid, leaving its mouthparts inserted like a pipette in the phloem, allows analysis of the sieve tube contents. Sieve tubes are found to contain a solution of amino acids and sucrose, the latter showing the expected diurnal variations in concentration (see 1 above).
5. The transport in the phloem is normally downwards. However, by darkening or removing the upper leaves, the sugars can be shown, by ringing, to move upwards in the phloem.

Contents of sieve tubes

It has been found using the aphid technique that sieve tubes contain a solution containing

1. up to 30% by weight sucrose
2. up to 1% by weight amino acids
3. traces of
   (a) sugar alcohols
   (b) ionic phosphate and potassium
   (c) hormones
   (d) viruses

Theories of phloem transport

Diffusion is too slow to account for the observed rates of transport; therefore, a number of theories have been proposed:

Mass flow hypothesis (Munch, 1930)

Material travels from a region of high concentration, e.g. photosynthesizing leaf chloroplast, to a region of low concentration, e.g. storage plastids in a root. In the region of high concentration water is taken up by osmosis and there is a difference in hydrostatic pressure between this region and the storage area.

This causes mass flow within the sieve tube lumina. The theory ignores the membrane barrier between the sieve tube and the plastid and assumes an empty sieve tube lumen and fully open sieve plate pores.

Transcellular strands (Thaine)

Thaine proposes that protein fibrils surrounding endoplasmic reticulum tubules pass from one end of the sieve tube to the other and has suggested that solutes pass along these fibrils due to the peristaltic action of the protein sheath, in a manner resembling cytoplasmic streaming. This is an active process.

Although the existence of these transcellular strands is not above doubt it is the only theory which accounts for transport of solutes in both directions in one sieve tube.

Electro-osmosis (Spanner)

Spanner suggests that the flow of nutrients is produced or maintained by electro-osmotic forces set up across the sieve plates (possibly a potassium ion gradient). This is theoretically possible but there is no direct evidence for it.

OSMOREGULATION AND EXCRETION

Underlying Principles

Primitive organisms probably arose in the sea. The tissue fluids of animals are dilute saline solutions with a composition similar to that of sea water. This is not mere coincidence: animals effectively bathe their tissues in sea water. Some organisms have an internal osmotic pressure that is isotonic with sea water.

Even this can create osmotic problems because as a consequence of carrying out metabolic activity the internal concentration of the animal will fluctuate from that of the medium in which it lives. Some marine animals moved up estuaries into fresh water.

As the external medium was then hypotonic to the animal cells, water flooded in by osmosis. A number of mechanisms to overcome these problems evolved:

1. The permeability of cell membranes to water and salt became altered.
2. The cells became enveloped in a rigid wall that prevented expansion.
3. The dilution was tolerated.
4. The water was pumped out by some means.

One or more of these methods was employed by organisms in fresh water; in all cases the problem was reduced by allowing their own internal concentrations to fall, thereby reducing the tendency of water to enter by osmosis. Some of these organisms then returned to salt water.

They then had an external medium that was hypertonic to their cells and water was lost by osmosis. Two methods have evolved to overcome this. In marine teleost fish salt water is drunk and the salts are eliminated by special glands and the kidney.

In marine elasmobranchs high urea and amino acid concentrations of the blood are tolerated so that the internal concentration of these fish is slightly hypertonic to sea water. When organisms moved on to land water retention became essential. This was achieved by:

1. Anatomical adaptations, e.g. waterproof body coverings.
2. Physiological adaptations, e.g. toleration of dehydration, mechanisms for controlling water balance.
3. Behavioural mechanisms, e.g. avoidance of dry areas and by aestivation during dry seasons.

During metabolic activities wastes are produced which are toxic and if allowed to accumulate would poison the organism. Many are removed by diffusion as part of some other process, e.g. carbon dioxide diffuses into the lungs during breathing.

In animals, however, nitrogenous wastes resulting from the breakdown of excess amino acids pose particular problems. The ammonia produced is especially toxic. If water is readily available it can be diluted sufficiently and removed. Where water needs to be conserved, e.g. in terrestrial organisms and marine vertebrates, nitrogenous wastes need to be more concentrated before being removed. They must be made less toxic before conversion to urea.

Where water is particularly scarce or flight makes the storage of watery urine impractical, the ammonia is converted to uric acid which requires almost no water for its removal. Birds and insects excrete uric acid. By virtue of their autotrophic mode of nutrition plants take in the materials they need and no more.

They therefore have almost no complex excretory products-only simple diffusible ones. Their high surface area/volume ratio is suited to the removal of these simple substances, e.g. carbon dioxide (dark), oxygen (light). Any complex excretory material can be removed with the leaves when they are discarded.

Points of Perspective

It would be an advantage to have dissected or seen a dissection of the urinary system of a small mammal. Experimental observations of osmoregulation in protozoa under different conditions would also be of benefit.

A good candidate will know the major differences in the composition of urine and the glomerular filtrate in a mammal, and if possible be able to quote figures. A range of anatomical, physiological and behavioural adaptations to preventing water loss in terrestrial organisms would be helpful.

Essential Information

Excretion is the separation and elimination of metabolic wastes usually in aqueous solution.

Excretory products in plants

1. Carbon dioxide—when the rate of respiration exceeds the rate of photosynthesis.
2. Oxygen—when the rate of photosynthesis exceeds the rate of respiration.

In plants there is no regular excretion of nitrogenous or other complex materials. Wastes such as silicates and tannins are moved to and stored in parts of the plant which will be removed, e.g. leaves and fruits.

Excretory products in animals

1. Carbon dioxide—from respiration
2. Excess water and mineral salts
3. Bile pigments
4. Nitrogenous substances, such as:

(a) *Ammonia* is highly toxic and never allowed to accumulate in living cells. It is soluble, readily diffusible and excreted in dilute form by freshwater animals and marine invertebrates. Animals excreting ammonia are said to be ammoniotelic.

(b) *Urea* is less toxic than ammonia but still needs to be diluted for elimination from the body. It is excreted by some terrestrial organisms and marine ones whose body fluids are hypotonic to sea water. These animals are ureotelic. Urea is formed from ammonia by the ornithine cycle.

(c) *Uric acid* is non-toxic. As it is highly insoluble, little water need be lost in its elimination from the body. It is a common excretory product of animals living under arid conditions. Insects and birds excrete uric acid and are said to be uricotelic.

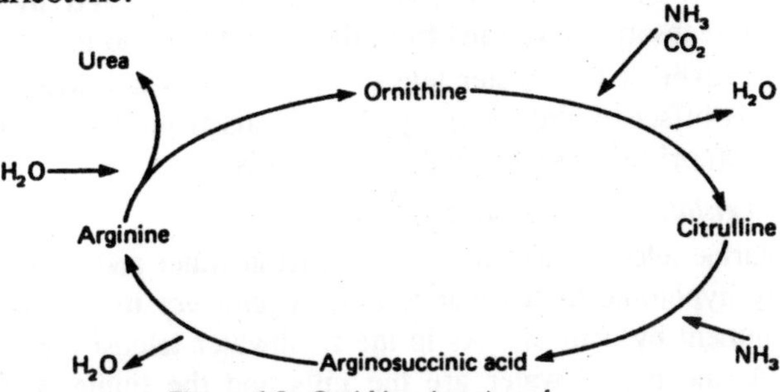

Figure 6.9: Ornithine (urea) cycle.

Ornithine (urea) Cycle

Excess amino acids cannot be stored. By demmination the amine group is removed from them resulting in the formation of ammonia. Ammonia is also toxic and so it is converted through a series of reactions, the ornithine cycle, to urea.

The enzyme arginase catalyses the formation of urea from arginine. The purpose of the cycle is to regenerate arginine using excretory ammonia. In mammals this cycle occurs in the liver cells.

Osmoregulation in freshwater protozoa

Freshwater protozoa have cell contents which are hypertonic to the medium in which they live. Since the cell membrane is semi-permeable water enters by osmosis. To prevent the animal bursting the excess water must be removed.

Freshwater protozoa such as *Amoeba* and *Paramecium* have contractile vacuoles which constantly fill and empty at the surface. The water which enters the cytoplasm must move into the contractile vacuole against the concentration gradient.

As the process requires energy, there are often aggregations of mitochondria near contractile vacuoles. The contractile vacuoles of freshwater protozoa moved to a medium with a higher salt concentration fill and empty less frequently than in fresh water.

Osmoregulation in freshwater teleosts

Although much of the surface of a teleost fish is impermeable to water there is still a tendency for water to enter across the gills and lining of the buccal cavity and pharynx. The organ of osmoregulation in the teleost is the kidney, which reacts to the influx of water in two main ways:

1. Production of a large volume of urine—aided by the many large glomeruli in the kidney.
2. Production of very dilute urine—achieved by extensive reabsorption of salts from the renal fluid into the blood stream.

However, a freshwater teleost still suffers some loss of salts and so this is offset by the active uptake of salts from the environment by special chloride secreting cells in the gills.

Osmoregulation in marine teleosts

Marine teleosts, which evolved in fresh water and have body fluids slightly hypotonic to sea water, have a *tendency* to lose water to the environment by osmosis. As in the freshwater teleost the areas of the body permeable to water are the gills and the lining of the *buccal cavity* and *pharynx*. To combat *dehydration* the *kidney*:

1. Produces a small volume of urine—there are relatively few and small glomeruli.
2. Produces the *nitrogenous* excretory product *trimethylamine* oxide. Freshwater teleosts excrete ammonia since they have plenty of water available for its dilution. *Trimethylamine oxide* is non-toxic and therefore a more suitable excretory product for marine teleosts since little water is needed for its *expulsion*.

Marine teleosts also drink sea water; the excess salts ingested are actively removed from the body by the chloride-secreting cells of the gills.

Osmoregulation in a marine elasmobranch

Marine elasmobranchs retain urea so that the body fluids are slightly hypertonic to sea water. There is consequently a slight influx of water but the excess is easily removed by the kidney. Elasmobranch tissues are unusual in being tolerant of high levels of urea.

Normally a high concentration breaks the hydrogen bonds in protein molecules, thus altering their shape and disrupting enzymatic properties. The ability to retain urea dispenses with the need to drink sea water and actively eliminate salts.

Osmoregulation in a terrestrial insect

All terrestrial organisms are liable to water loss through evaporation. In an insect the loss is reduced in three main ways:

1. Having an impermeable surface—the cuticle is coated with wax. Retention of water can never be complete and there is bound to be some loss of water from respiratory surfaces.
2. Production of uric acid - because this nitrogenous waste product is non-toxic and insoluble, it can be eliminated from the body as a semi-solid without the loss of water.
3. Reabsorption of water by the Malpighian tubules and rectal gland. The Malpighian tubules are a bunch of blind-ending tubules at the junction of the mid-gut and rectum. Potassium urate is produced by insect tissues and this enters the Malpighian tubules with water and carbon dioxide.

   In the tubules they form uric acid and potassium bicarbonate. The potassium bicarbonate is reabsorbed into the haemolymph. Reabsorption of water from the uric acid occurs in the Malpighian tubules and the rectal gland so that the urine eliminated is in a very concentrated form.

Mammalian Kidney

The basic unit of nitrogenous excretion in a mammal is the kidney tubule, or nephron. Each human kidney contains over one million nephrons and filters about 120 cm^3/min. Over the length of the nephron reabsorption from the glomerular filtrate takes place and results in the formation of urine which is hypertonic to the blood.

Functions of the mammalian kidney

1. Excretion of metabolic wastes, especially urea
2. Osmoregulation
3. Maintaining the acid-base balance of the body. This is mainly

brought about by the excretion or retention of H^+ and HCO_3^-; the pH of the urine can vary between 4.5 and 8.0

4. Maintaining the balance of ions, e.g. Na^+, K^+, Ca^{2+}, Mg^{2+}, H^+, Cl^-, HCO_3

Formation of hypertonic urine

The loop of Henle, found in birds and mammals, constitutes a counter-current multiplier, the active mechanism being the transport of sodium ions from the ascending to the descending limb. This movement of sodium ions results in a very high concentration developing in and around the apex of the loop of Henle, deep in the medulla.

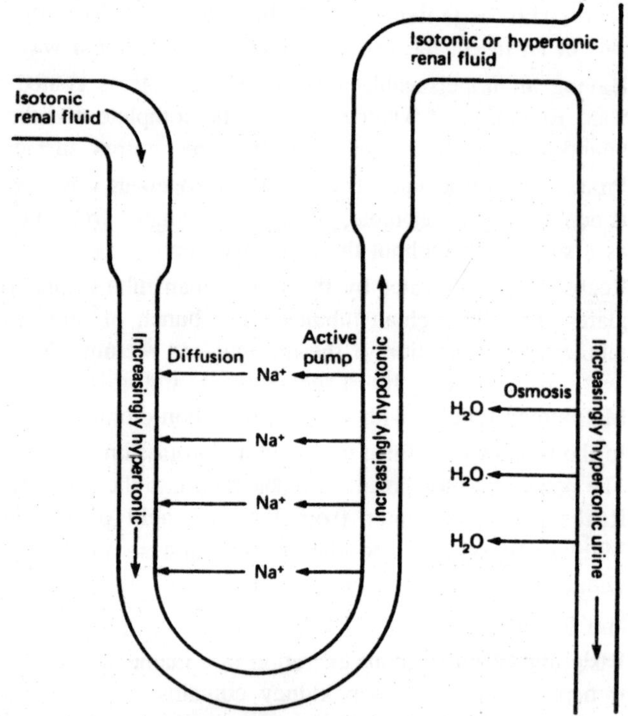

Figure 6.10: Counter-current multiplier.

In this region the thin-walled collecting ducts open into the renal pelvis. The high concentration of sodium in the tissues causes water to be drawn out of the collecting ducts by osmosis. Consequently the renal fluid which has been isotonic or hypotonic to the blood through the length of the nephron becomes hypertonic in the collecting ducts. It has been estimated that 99% of the fluid filtered by the Bowman's capsule is reabsorbed by the nephron.

The length of the loop of Henle is proportional to the concentration of urine, e.g. desert mammals, such as the kangaroo rat, which produce very hypertonic urine have very long loops of Henle.

Control of water and salt balance in mammals

Antidiuretic hormone (ADH)

The permeability of the collecting ducts to water is affected by the hormone ADH, secreted by the posterior lobe of the pituitary. The presence of ADH increases the permeability of the membranes and results in the production of more hypertonic urine. If no ADH is present the membrane permeability is lower and more dilute urine is produced.

The production of ADH is controlled by the hypothalamus whose osmoreceptors monitor the solute concentration of the blood. If the solute concentration of the blood is high ADH secretion is stimulated, the permeability of the membranes of the collecting ducts increases, water is reabsorbed into the blood and very hypertonic urine is produced.

ADH secretion may also be triggered, via the hypothalamus, by blood volume receptors in the walls of the heart, the aorta and the carotid arteries.

Aldosterone

This hormone produced by the adrenal cortex stimulates the reabsorption of sodium from the kidney. When the sodium concentration in the kidney tubule decreases the kidney produces the enzyme renin which catalyses the conversion of the blood protein angiotensinogen into angiotensin. Angiotensin stimulates the adrenal cortex to produce more aldosterone.

7

FEEDING AND NUTRITION

AUTOTROPHIC NUTRITION (PHOTOSYNTHESIS)

Underlying Principles

One method of distinguishing living material from non-living is to compare the way they relate to the Second Law of Thermodynamics. This law states that all matter tends to high entropy (i.e. it tends to become disordered). This is true of non-living material which tends towards a state where it possesses the minimum of energy and becomes disordered.

Living organisms on the other hand are highly ordered systems possessing much stored energy. In fact even living organisms tend to lose energy and become disordered. Where they really differ from non-living material is in their ability to replace the lost energy from outside themselves. In animals and the fungi this energy is obtained from food, which comes ultimately from green piants.

The green plants obtain energy from the earth's major source, the sun, by using light to produce complex organic molecules from simple inorganic ones in the process of photosynthesis. All life is directly or indirectly dependent on this the most fundamental process in living organisms.

Points of Perspective

A good candidate would have a knowledge of alternative mechanisms and pathways in autotrophic organisms. Examples include the C_4 pathway and a range of chemosynthetic mechanisms where inorganic substances are used to provide energy rather than light, e.g. oxidation of ammonium compounds to nitrites by *Nitrosomonas.*

It would be useful to have some appreciation that photosynthesis can occur in some prokaryotic cells without chlorophyll, e.g. in the purple sulphur bacteria.

Some information on the inter-relationship between the photosynthetic pathway and other biochemical pathways in a plant cell would be helpful, in particular how the many other substances in a cell are formed from the products of photosynthesis.

Essential Information

Light

Visible light represents that part of the electromagnetic radiation spectrum which lies between 380 nm (violet) and 750 nm (far red). Three properties of light that are of importance to organisms are:

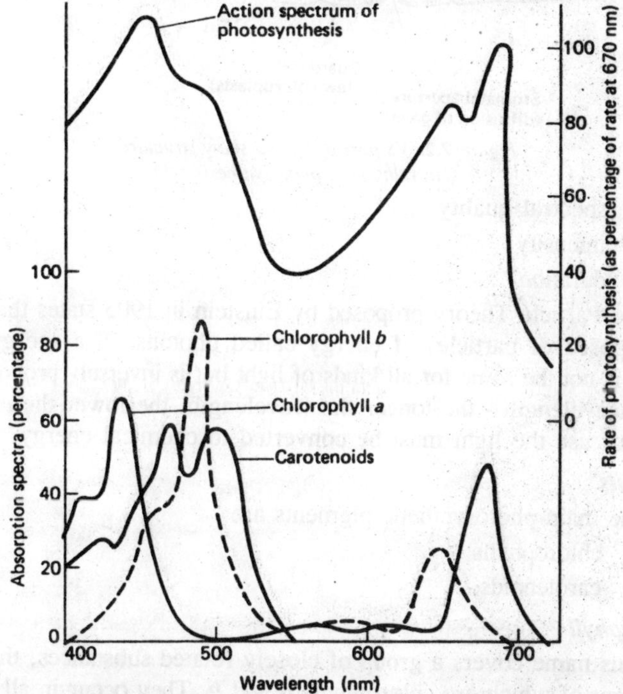

Figure 7.1: Graph to show relationship between action spectrum for photosynthesis and absorption spectra of major photosynthetic pigments

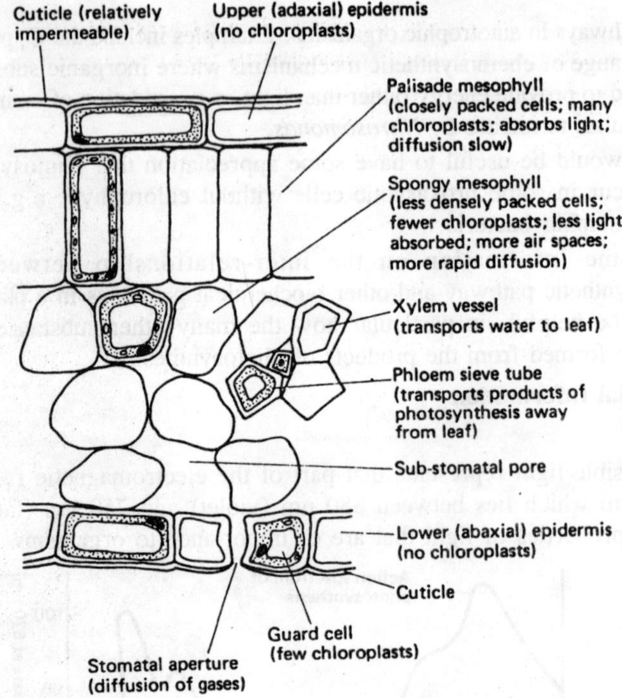

Figure 7.2: VS part of leaf to show structure in relation to photosynthesis

1. spectral quality
2. intensity
3. duration.

The Particle Theory proposed by Einstein in 1905 states that light is composed of particles of energy called photons. The energy of a photon is not the same for all kinds of light but is inversely proportional to the wavelength - the longer the wavelength, the lower the energy. To be of use the light must be converted to chemical energy.

Pigments

The main photosynthetic pigments are:

1. chlorophylls
2. carotenoids.

Chlorophyll

This name covers a group of closely related substances, the most important of which are chlorophylls *a* and *b*. They occur in all higher plants in the approximate ratio of 2 : 1.

Chlorophyll *a* $C_{55}H_{72}O_5N_4Mg$ Chlorophyll *b* $C_{55}H_{70}O_6N_4Mg$

Chlorophyll is usually contained within chloroplasts. The total weight of chlorophyll in green leaves varies between 0.55-0.20% of the fresh weight.

Carotenoids

These are a large group of hydrocarbon pigments found throughout the plant. About 100 different carotenoids have been recognized. There are two major groups: the carotenes and xanthophylls.

They are important pigments in photosynthesis, especially in some groups of algae, e.g. fucoxanthin in the brown algae (Phaeophyceae). They strongly absorb light in the violet end of the spectrum.

Mechanism of photosynthesis

In 1932 Emerson and Arnold exposed plants to flashes of light lasting 10^{-4} seconds and varied the period of dark between the flashes. They found that the yield of carbohydrate increased as the length of the dark periods increased up to a maximum dark period of 20 milliseconds (0.02 s) at 25°C.

If the temperature was reduced so was the yield of carbohydrate, although this could be compensated for by increasing the period of darkness. Increasing the duration of the flashes of light had no effect. These experiments suggest that photosynthesis has two stages:

1. A light stage (photochemical stage) which requires light, but is unaffected by temperature.
2. A dark stage (chemical stage) which does not require light and is affected by temperature.

The Light Stage

When light is received by a chlorophyll molecule one of its electrons is lost, leaving the chlorophyll molecule positively charged and chemically less stable. This electron may return to the chlorophyll molecule, via the carrier system, thus stabilizing it again.

In doing so some of the energy it loses is used to combine adenosine diphosphate and inorganic phosphate into adenosine triphosphate (phosphorylation). As further light can again raise the electron's energy and the process be repeated, it is termed *cyclic photophosphorylation.*

The electron may however not return directly to the chlorophyll molecule. It may combine with hydrogen ions (H^+) that result from the natural dissociation of water.

$$\underset{\text{water}}{H_2O} \rightleftarrows \underset{\text{hydrogen ion}}{H^+} + \underset{\text{hydroxyl ion}}{OH^-}$$

In doing so a hydrogen atom is formed which is immediately taken up by a hydrogen acceptor such as ***nicotinamide adenine dinucleotide phosphate*** (NADP) which enters the dark reaction. The stability of the chlorophyll is restored in this case by the hydroxyl ion (OH^-) from the dissociation of water donating its extra electron to the chlorophyll molecule.

The electron is transported via an electron carrier system and ATP is yielded as before. The resulting OH is combined with others to form water and oxygen, the latter being evolved as oxygen gas.

$$4(OH) \rightleftarrows 2H_2O + O_2 \uparrow$$

Since a different electron is returned to the chlorophyll molecule the process is termed *non-cyclic photophosphorylation*.

The dark stage (Calvin Cycle)

The reduced nicotinamide adenine dinucleotide phosphate ($NADPH_2$) from the light reaction is used to reduce carbon dioxide using the ATP formed in the light reaction as the source of energy. Carbon dioxide absorbed by the plant is combined with a five carbon substance *ribulose diphospate* and the resulting unstable intermediate immediately splits to give two molecules of the three carbon substance *phosphoglyceric acid* (PGA).

Most of this PGA is used to reform ribulose diphosphate but some is reduced by the $NADPH_2$ to give triose phosphate, which in turn is built into glucose phosphate which polymerizes into starch by condensation.

C_4 photosynthetic pathway

Plants such as sugar cane and maize have evolved an alternative mechanism for the fixation of carbon dioxide that involves the use of *phosphoenol pyruvic acid* (PEP) instead of ribulose diphosphate. The first formed product of this process is the four carbon oxaloacetic acid instead of PGA.

It provides a means of obtaining carbon dioxide and storing it chemically for later conversion to PGA, which is of particular advantage in hot tropical climates where carbon dioxide levels may be low during the day.

Importance of Photosynthesis

1. Releases oxygen required by animals and plants for aerobic respiration.
2. Provides a store of useful chemical energy by converting inorganic substances to stable organic substances of high potential

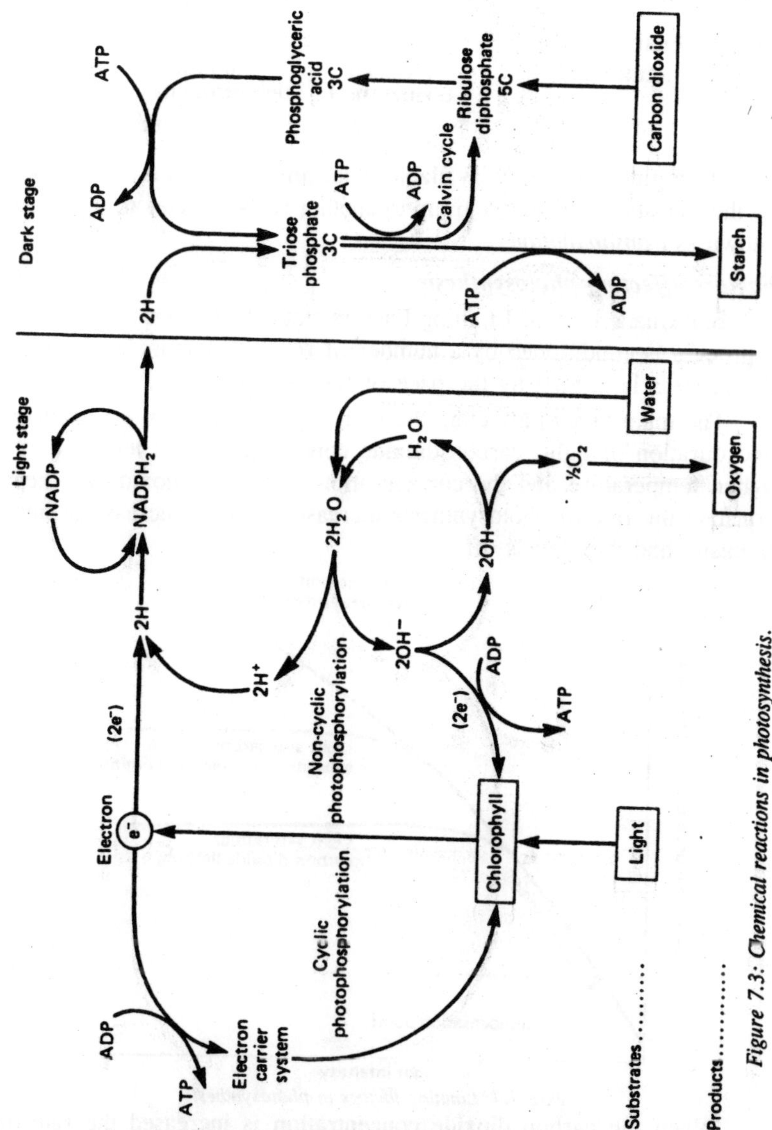

Figure 7.3: Chemical reactions in photosynthesis.

chemical energy. It has been estimated that plants annually fix 35×10^{15} kg of carbon.

Synthesis of nitrogenous compounds

The nitrogen of ammonia is brought into organic combination mainly by the formation of glutamic acid

$$\alpha \text{ ketoglutaric acid} + \underset{\text{ammonia}}{NH_3} + \underset{\text{reduced coenzyme I}}{CoI - H_2} \xrightleftharpoons[\text{dehydrogenase}]{\text{glutamic}} \text{glutamic acid} + CoI$$

Once glutamic acid is available other amino acids may be formed by the transfer of its amino group to another carbon skeleton, a process known as *transamination*.

Factors affecting photosynthesis

Blackmans Law of Limiting Factors states that when the speed of a process is conditioned by a number of separate factors, the rate of the process is limited by the pace of the slowest factor.

The main factors affecting the rate of photosynthesis are intensity and duration of light, carbon dioxide concentration and temperature. When temperature and the concentration of carbon dioxide are kept constant the rate of photosynthesis increases with an increase in light intensity and then levels off.

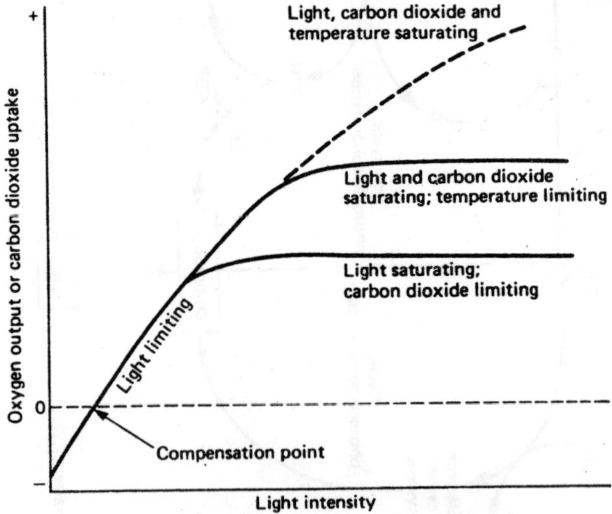

Figure 7.4: Limiting factors in photosynthesis.

When the carbon dioxide concentration is increased the rate of photosynthesis increases to a maximum before levelling out. If the temperature is then increased the rate of photosynthesis steadily rises instead of levelling off.

Compensation point

This is the point at which the rate of carbon dioxide production

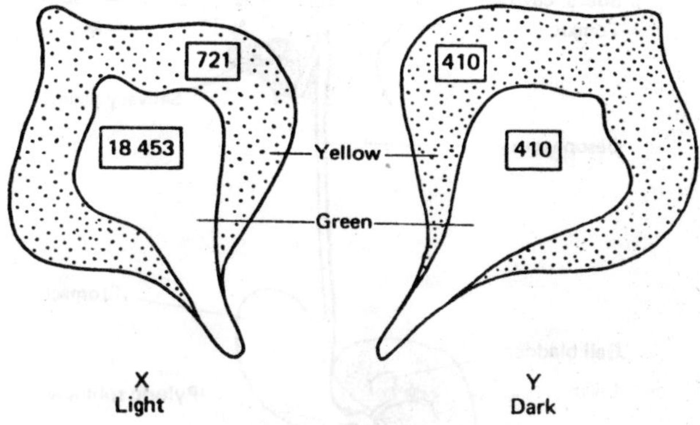

Figure 7.5: Experiment to investigate photosynthesis in variegated leaves.

from respiration is compensated exactly by the rate of carbon dioxide uptake in photosynthesis. Below the compensation point the rate of photosynthesis is less than the rate of respiration and so carbon dioxide is evolved.

Above the compensation point the rate of photosynthesis exceeds the rate of respiration and so oxygen is evolved. At the compensation point there is no net exchange of gases.

HETEROTROPHIC NUTRITION (HOLOZOIC)

Underlying Principles

The evolution of autotrophic organisms inevitably led to the development of organisms that obtained their energy not by photosynthesis but by consuming the autotrophs. These were the herbivores. They developed various mechanisms for ingesting the food and digesting the cellulose walls that surrounded the plant cells.

In turn there evolved organisms that obtained energy by consuming the herbivores. These were the carnivores. They developed a variety of mechanisms for capturing prey organisms as well as for ingesting them. Unlike autotrophs the energy source of heterotrophs is in complex form and contains some unwanted materials.

The food must therefore be digested, the required materials absorbed and the unwanted ones eliminated. This digestion is carried out by enzymes each with an optimum pH. For enzymes to be efficient the food they act upon must have a large surface area.

In addition to enzymes the digestive system therefore also produces

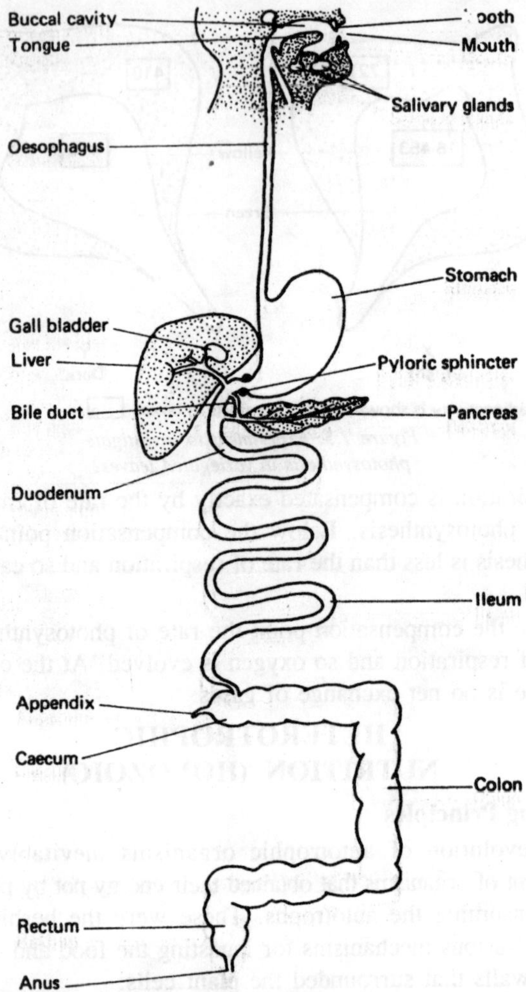

Figure 7.6: Human digestive system.

substances to adjust the pH, water as a medium for the enzymes to act in and substances to increase the surface area of the food. To speed the absorption of the digested food there are methods to increase the surface area of the absorbing surface. To maintain a diffusion gradient this surface is well supplied with blood vessels and the food is kept moving. The surface is thin and moist.

Points of Perspective

Candidates should have carried out a full dissection of the digestive

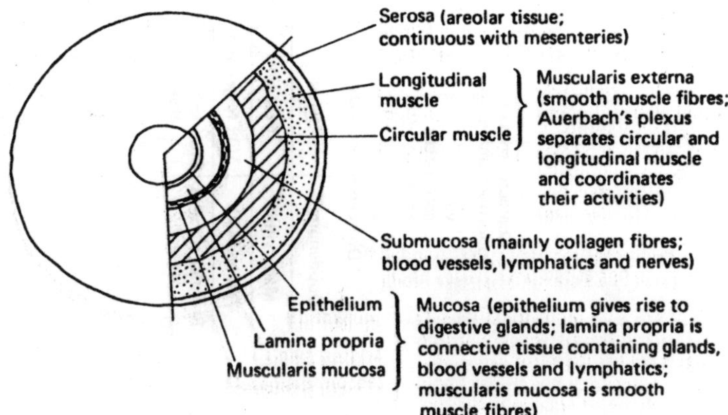

Figure 7.7: General plan of the alimentary canal.

system of a mammal and be able to recognize all the major parts. It would be beneficial to also have carried out a number of experiments on enzymes to determine the effects of pH, temperature, inhibitors and other factors on the rate of action of digestive enzymes such as pepsin, trypsin, amylase and lipase.

A knowledge of a wide range of feeding mechanisms with specific examples would be useful. An understanding of the precise role of each food substance in the body and the consequent symptoms of its deficiency would also be helpful.

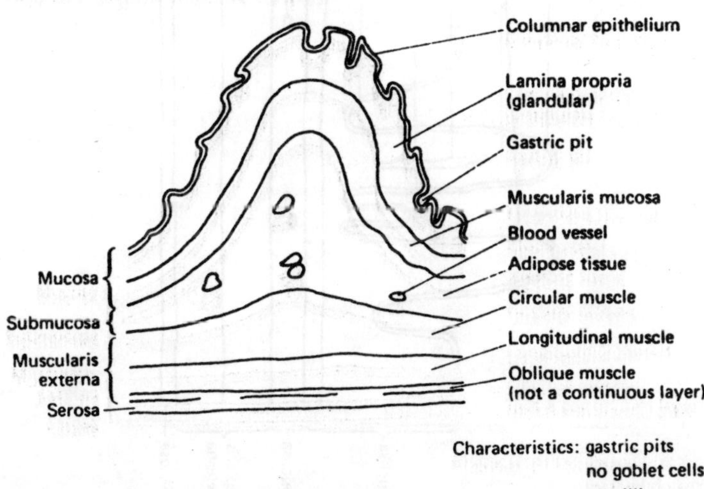

Figure 7.8: LS stomach.

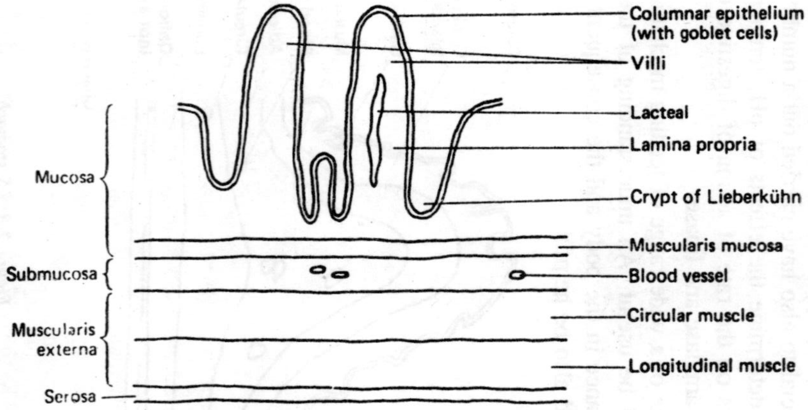

Figure 7.9: LS ileum.

Characteristics of duodenum and ileum:

villi
goblet cells
crypts

Differences between

Duodenum	Ileum
Many villi	Few villi
Villi leaf-like	Villi finger-like
Brunner's glands present	Brunner's glands absent

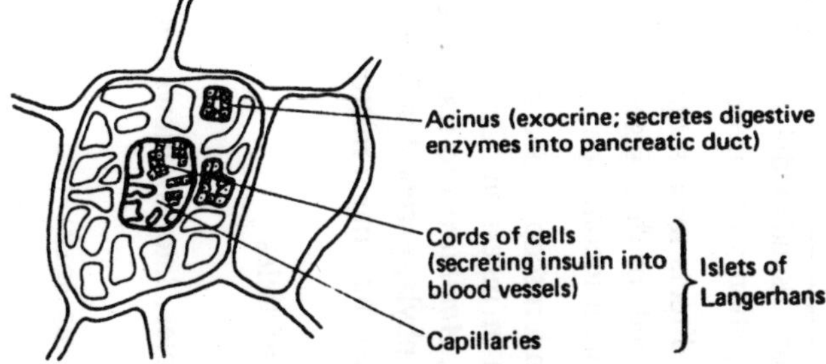

Figure 7.10: TS pancreas.

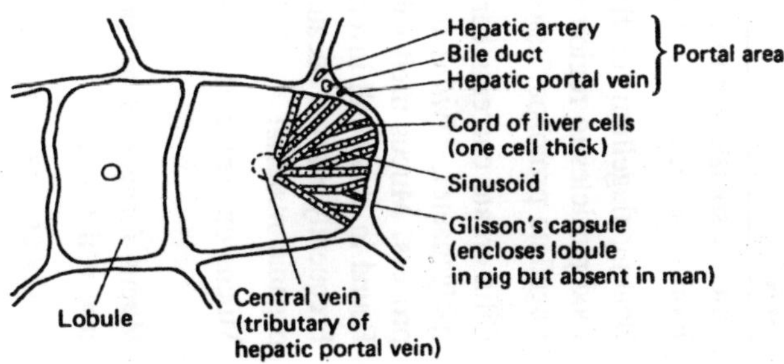

Figure 7.11: TS liver.

Essential Information

Structure and histology of human digestive system. Heterotrophic organisms consume complex food material. This food must be

1. obtained (may involve movements to capture or find new food sources)
2. ingested (feeding mechanisms)
3. physically digested (e.g. by teeth, radula, gizzard)
4. chemically digested (largely by enzymes)
5. absorbed (the required material)
6. eliminated (the unwanted material).

Physical Digestion

To make food manageable for movement along the alimentary canal and give it a larger surface area for enzymes to act on, different mechanisms for physically breaking up the food have evolved.

The structures involved include a gizzard (earthworms, birds), a radula (gastropod molluscs), a gastric mill (crayfish), mandibles (insects) and teeth (vertebrates).

The teeth in particular are well adapted to the diet. Indeed, not

Table 7.1: Feeding mechanisms in animals (based on the classification by Yonge).

Food type	*Mechanism*	*Examples*	*Notes*
Small particles	Pseudopodial	*Amoeba*	Often associated with locomotion. Pseudopodia enclose food in a vacuole in which digestion takes place
	Flagellate	*Euglena*, sponges	Beating flagellum (or flagella) directs microscopic food particles to region specialized for their ingestion. In flagellate protozoa, associated with locomotion
	Ciliary	*Paramecium, Sabella*	Cilia create currents carrying microscopic food particles (polychaete), *Mytilus* to region of ingestion. Continuous process. Mucus used (bivalve), *Amphioxus* to trap food. Found in almost all invertebrate phyla, except Arthropods. Frequently marine because sea contains much microscopic organic material
	Tentacular	Sea cucumber (echinoderm)	Mucus on tentacles traps food particles
	Mucoid	*Vermetus (mollusc)*	Mucus forms veil to trap particles, later swallowed and new veil formed
	Setous	*Daphnia*, some aquatic insect larvae e.g. *Culex*	Chitinous setae on appendages trap small food particles. May be associated with locomotion
	Other	Basking shark, herring,	Planktonic organisms filtered out of water by various

(Table 7.1 Contd.)

Food type	Mechanism	Examples	Notes
		flamingo, whalebone whales	mechanisms. Basking sharks and herrings use gills to filter particles out of respiratory currents; flamingo uses beak for blue-green algae; whales feed on krill using keratinous plates
Large particles	Swallowing inactive food	*Lumbricus, Arenicola*	Non-selective swallowing of mud, silt, sand etc., the organic content supplying nourishment
	Scraping and boring	*Helix, Littorina, Teredo* (molluscs), caterpillars, termites	*Molluscs* usually use radula, insects mandibles. *Teredo* uses shell valves. Herbivores and carnivores
	Seizing and swallowing	Coelenterates, *Nereis,* dogfish, snakes, birds, seals	No chemical or mechanical breakdown before swallowing. Prey may be paralysed. Swallowing aided by mucus secretions
	Seize and masticate before swallowing	Squid, crabs, dragonfly, many mammals	Involves specialized organs for reducing size of food particles prior to digestion e.g. beak of squid, chelae and mandibles of crab; maxillae and labium of dragonfly, teeth of mammals
	Seize and digest externally before swallowing	Starfish, spiders, blowfly larvae	Enzymatic digestive juices secreted over, or injected into, food mass to make it small enough to swallow. In spiders, pedipalps inject proteolytic enzyme into prey
Fluids or	Piercing and	Leeches, hookworms,	Mouth parts modified for piercing and also sucking

(Table 7.1 Contd.)

Food type	*Mechanism*	*Examples*	*Notes*
soft tissues	sucking	female mosquitoes, aphids, fleas, lice, vampire bats	devices developed. Often feed on plant nectar and sap but may feed on blood. Often ectoparasites (hookworms are endoparasites). Blood feeders have anticoagulant in saliva. May be vectors of endoparasitic organisms
	Sucking only	Nematodes, lepidoptera, male mosquitoes, housefly	Often live on plant secretions like nectar or as endoparasites surrounded by their food source. Simple suctorial action only
	Absorb through general body surface	*Trypanosoma*, *Monocystis*, cestoda	Only in animals permanently bathed in fluid nutrient medium. No real feeding mechanisms and usually no alimentary canal. Generally endoparasitic

only teeth, but the whole digestive system has become modified according to whether the organism is herbivorous or carnivorous.

Table 7.2: The main differences between herbivorous and carnivorous mammals.

Herbivores	*Carnivores*
Open pulp cavity in teeth	Closed pulp cavity in teeth
Incisors for cutting or gnawing food or absent e.g.	Incisors pointed for nipping and biting upper jaw of sheep
Canines small or absent leaving gap called diastema e.g. in rabbit	Canines very well developed for piercing and tearing
Carnassial teeth absent	Carnassial teeth present for shearing flesh
Cheek teeth flattened with ridges of enamel and grooves of dentine. Adapted for grinding	Cheek teeth pointed (cusps) and adapted for shearing
Upper teeth meet lower teeth to facilitate grinding	Upper jaw wider than lower to facilitate shearing action
Articulation of lower jaw permits lateral movement	Articulation of lower jaw prevents lateral movement and therefore reduces the danger of dislocation
Skull smooth without well-developed processes for muscle attachment	Coronoid process, zygomatic arch and sagittal crests well developed for the attachment of large powerful muscles
Cellulose digestion carried out by micro-organisms in rumen and reticulum (ruminants) or caecum and appendix	No cellulose in diet therefore no rumen; caecum and appendix are small
Relatively long alimentary canal to digest large volumes of vegetation	Relatively short alimentary canal because diet rich in protein with relatively little indigestible material

Chemical Digestion

Chemical digestion is carried out primarily by enzymes with the aid of many other substances that help to break up the food (e.g. bile salts) and adjust the pH to the optimum for particular enzymes.

Table 7.3: Summary of digestion.

Organ	*Secretion*	*pH*	*Site of action*	*Production induced by*	*Secretion (including enzyme)*	*Effect*
Salivary glands	Saliva	Slightly alkaline	Mouth	Visual/olfactory expectation; reflex stimulation	Salivary amylase	Starch → maltose via dextrin
					Mucin	Sticks bolus
					Salts	Correct pH
Gastric glands of stomaeh	Gastric juice	Very acid	Stomach	Gastrin (hormone)	Hydrochloric acid	Provides correct pH
				Reflex stimulation	Pepsin	Protein → polypeptides
					Rennin	Caseinogen → casein (chymase)
					Mucus	Lubrication; prevents autolysis
Liver	Bile	Alkaline	Duodenum	Cholecystokinin induces release of bile	Bile salts	Emulsify fats
					Bile pigments	Excretory products of haemoglobin breakdown
				Reflex action		
				Secretin causes production of bile	Sodium hydrogen carbonate	Correct pH
					Cholesterol	Excretory
Pancreas	Pancreatic	Alkaline	Duodenum	Secretin induces	Trypsin	Protein → peptides and amino acids

(Table 7.3 Contd.)

Organ	*Secretion*	*pH*	*Site of action*	*Production induced by*	*Secretion (including enzyme)*	*Effect*
	juice			production of pancreatic juice	Carboxypeptidase	Peptides → amino acids
				Pancreozymin induces release of pancreatic juice	Chymotrypsm	Casein → peptides + amino acids
					Amylase	Starch → maltose
					Lipase	Fats → fatty acids and glycerol
					Nucleases	Nucleic acids → nucleotides
Small intestine wall	Intestinal juice	Alkaline	Small intestine	Mechanical stimulation of intestinal lining	Enterokinase	Trypsinogen → trypsin
					Aminopeptidase	Peptides → amino acids
					Amylase	Starch → maltose
					Maltase	Maltose → glucose
					Sucrase	Sucrose → glucose and fructose
					Lactase	Lactose → glucose and galactose
					Nucleotidases	Nucleotides → organic bases, pentose, sugar and phosphoric acid

Table 7.4: Absorption and fate of major digestive products.

End product of digestion	*Form in which absorbed*	*Where absorbed*	*Mechanism of absorption*	*Fate of products*	*Storage*	*Use*
Monosaccharides	Possibly as a complex e.g. phosphate sugar	Capillaries of villi	Diffusion and active transport	Most remain in general circulation. Excess converted to glycogen in the liver	Glycogen in liver	Respiratory substrate
Fatty acids and glycerol	Fatty acids and glycerol; some as droplets of neutral fat (chylomicrons)	Mainly lacteals in villi; some into capillaries	Diffusion and active transport	Fat enters blood stream from lymphatic system (in neck)	Fat under skin, in mesenteries etc.	Respiratory substrate; insulation; protection; phospholipids (structural)
Amino acids	Amino acids	Capillaries of villi	Diffusion and active transport	Deamination in liver: nitrogenous residues converted to urea; carbon residues mainly in general circulation but some used for carbohydrate or fat	After deamination as fat or carbo-hydrate	Protein synthesis; deaminated proteins may become part of carbohydrates or fats synthesis

Absorption

The soluble products of digestion are absorbed both by diffusion and active transport. To aid these processes an absorbing surface such as the villi of the small intestine has the following features:

1. It is thin (usually a single-celled epithelial layer).
2. It is moist.
3. It is well supplied with blood vessels (to maintain a diffusion gradient by removing the absorbed material).
4. It has a large surface area (for more rapid absorption).

The absorbed materials are transported to various organs of the body and either utilized immediately or stored for later use.

Elimination

Unwanted food is eliminated by periodic egestion from the organism, a process known as defaecation. It is important to note that, as such material has never cro ssed an epithelial boundary or been involved in the organism's metabolism, the process is elimination (egestion) and not excretion.

Interconversion

Although each absorbed product of digestion has a particular role in the body, a temporary shortage of one product can be compensated for by conversion of other readily available products.

HETEROTROPHIC NUTRITION (SAPROTROPHS AND PARASITES)

Underlying Principles

When autotrophic and holozoic organisms die, their remains comprise complex chemicals which may be broken down to simpler forms. The energy released in this process is utilized by groups of organisms called saprotrophs, which speed up the breakdown of the dead and decaying remains.

The energy released is used to build up their own complex structures. In addition saprotrophs play a vital role in releasing valuable elements such as carbon, nitrogen and phosphorus in a form in which they can be absorbed by autotrophs and so he recycled.

It is possible that such organisms arose when an autotrophic organism became unable to photosynthesize, e.g. lost the ability to make chlorophyll. If it were to survive it would need to feed heterotrophically.

With no means of obtaining and ingesting complex food it would need to live on dead organisms and digest them externally, absorbing the soluble products.

In a similar way any organism that lost, by mutation, the ability to obtain or utilize a particular substance, would need to obtain it from some external source. One way would be to obtain it from another living organism. This is the basis of parasitism.

In due course the parasite becomes increasingly dependent on its host for other nutrients. It is not in the parasite's interest to kill the host, indeed the less obvious it makes its presence the more likely it is to survive.

The host and parasite may develop such a close relationship that they become mutually beneficial. This is symbiosis.

Points of Perspective

It would be an advantage to have a wide knowledge of the economic importance of saprophytes and parasites. An understanding of the various methods of controlling important human parasites would be beneficial.

A good candidate will understand the problems of defining terms such as parasitism, saprotrophism, symbiosis, mutualism and commensalism, and appreciate that the divisions between the categories are not clear.

Essential Information

There are three types of heterotrophic nutrition:

1. holozoic
2. saprotrophic ('sapros' = rotten; 'trophe' = nourishment)
3. parasitic.

Saprotrophic nutrition

Saprotrophs are organisms obtaining their energy from the dead remains of other organisms. Saprotrophic plants are known as *saprophytes* and animals as *saprozoites*.

The importance of saprotrophs lies in their role in breaking down complex organic material and so releasing the valuable elements such as nitrogen, carbon and sulphur, usually in the form of an oxide, e.g. nitrate, carbon dioxide, sulphur dioxide, or in a reduced state, e.g. ammonia. methane. Many bacteria and fungi are saprotrophs, e.g. *Myxococcus, Mucor*.

Parasites

Parasitism is an association between two organisms whereby one (the parasite) derives some benefit and the other (the host) does not, and may in fact be harmed. However. the most well-adapted parasites inflict little damage on their hosts.

Plants are parasitized mainly by fungi and bacteria; they are poor hosts for invertebrates because:

1. They remain in one position and so it is more difficult for the parasite to reach a new host.
2. Their cellulose cell walls are difficult for animals to digest.
3. They have few internal cavities suitable for parasites.

The most successful invertebrate parasites of plants are nematodes. In animals the habitats most used are:

1. The body surface and its infoldings such as the buccal cavity, lungs, external gills and nostrils.
2. The alimentary canal and its associated organs such as the liver and bile duct.
3. Internal tissues such as blood and muscle.

Parasites using the body surface as a habitat are termed *ectoparasites* (ektos=outside) and those using internal habitats are *endoparasites (endon=within)*.

Most parasites are endoparasites because the inside of the body provides an environment which is rich in nutrients and not subject to extremes of environmental conditions.

Transmission of Parasites

The transmission of ectoparasites is relatively simple, e.g. by jumping in the flea *Pulex*. Since ectoparasites are subjected to varying conditions existence away from the host can be tolerated temporarily. The transmission of endoparasites however presents three main problems:

1. leaving the host
2. living away from the host
3. entering anew host.

Table 7.6: Methods used by parasites to escape from their hosts.

Habitat	*Example*	*Method*
Lungs	*Bacillus tubercle*	Through the air
Intestines	*Ascaris*	In the faeces
Intestines (larvae encysted	*Trichinella*	Disintegration of host tissues after death

in muscle)		
Blood	*Schistosoma*	Urine (eggs in wall of bladder)
Blood	*Plasmodium*	Intermediate blood sucking organism
Under skin	*Dracunculus*	Discharge larvae through skin of host

Living away from the host

There may be:

(a) dormant stages, e.g. *Taenia*

(b) free-living stages, e.g. *Haemonchus*

(c) intermediate hosts, e.g. *Fasciola.*

Entering a new host

The parasite may:

(a) be eaten, e.g. *Taenia*

(b) penetrate the skin or body surface of a new host after leaving an intermediate host, e.g. *Schistosoma*

(c) be injected by an intermediate host, e.g. transmission of *Plasmodium* by the mosquito.

The main features of parasites

1. A means of attachment to help the parasite maintain its position in/on the host, e.g. hooks, suckers.
2. Degeneration of certain unnecessary organ systems, e.g. the digestive system of gut parasites is reduced or absent because the parasite is surrounded by a large supply of predigested food. There are often special mechanisms developed to move a parasite from one host to another and so there is no need for a well developed locomotory system.
3. Protective devices are often developed to protect the parasite from the host's defences, e.g. schistosomes absorb glycolipids from the host on to their surface to imitate the host tissue and thus forestall immunological attack.
4. Penetrative agents such as cellulases in fungal plant parasites.
5. Vector or intermediate host, e.g. *Fasciola* (liver fluke of sheep), requires an intermediate host, the snail, for the completion of its life cycle. The blood-sucking female mosquito *Anopheles is* the intermediate host for *Plasmodium* (malarial parasite) and is thus the vector of the disease malaria.

6. Production of many eggs because of the difficulty of finding a new host, e.g. *Diphyllobothrium* (fish tapeworm) produces 36 million eggs a day.
7. Dormant or resistant phases to overcome the period spent away from the host, e.g. eggs in *Ascaris*, encysted cercariae in *Fasciola*.

Symbiosis

This term is most commonly applied to a relationship in which both organisms benefit, e.g. *Zoochlorella* inhabiting the endodermal cells of *Hydra viridissima*. In this relationship *Zoochlorella* obtains protection, shelter, carbon dioxide (for photosynthesis) and nitrogen (from excretory wastes) from *Hydra; Hydra* obtains oxygen and carbohydrates from the photosynthesis of *Zoochlorella*.

Lichens are a symbiotic relationship between an alga and a fungus in which the alga gains water, mineral salts and protection from desiccation and the fungus obtains oxygen and carbohydrates. The hermit crab *Pagurus* gains protection and camouflage from the sea anemone *Adamsia palliata* which lives on its shell. The sea anemone obtains food dropped by the hermit crab.

Alternative classification of relationships

In some schemes symbiosis is taken to mean 'living together' and this is subdivided as follows:

1. Mutualism: a relationship in which both organisms benefit.
2. Commensalism: a relationship in which one organism benefits and the other neither benefits nor suffers.
3. Parasitism: a relationship in which one organism benefits and the other is harmed.

8

EXCHANGE OF GASES

CELLULAR RESPIRATION

Underlying Principles

All energy exchanges whether in living or non-living systems are governed by the laws of thermodynamics. The first of these states that energy cannot be created or destroyed, only changed from one form to another.

The second states that natural processes proceed when there is an increase in the entropy (randomness) of a system. In effect this means that systems tend to lose their energy and assume their lowest possible energy level. Living organisms are highly ordered (have low entropy) which they achieve by the breakdown of glucose in the process called respiration.

In doing so they are increasing the entropy of their environment. The breakdown of glucose occurs in three main stages in aerobic respiration and one in anaerobic respiration.

In both cases the first stage is the splitting of the 6-carbon glucose molecule into two 3-carbon molecules (glycolysis).

In anaerobic respiration this is followed by one of a variety of reactions. In aerobic respiration the other stages are the tricarboxylic acid (Krebs') cycle and the electron carrier pathway.

All processes have evolved to allow the controlled release of the energy which is temporarily stored in chemical form as adenosine triphosphate (ATP).

Points of Perspective

It would be an advantage to understand the basic chemistry of oxidation and reduction reactions. Although examination boards do not require knowledge of the chemical formulae of intermediates in the biochemical processes of respiration, a good candidate will at least know the names of the major intermediates.

It would be helpful to know the inter-relationships between the respiratory pathway and other metabolic pathways, e.g. photosynthesis. In particular, it would be useful to know where fatty acids, glycerol and amino acids can be fed into the respiratory pathway.

Essential Information

The energy essential to maintain the, highly structured state of living organisms is derived from the food they manufacture (autotrophs) or consume (heterotrophs).

Whatever form the food initially takes it is converted into carbohydrate, usually the hexose sugar glucose, before being respired. Respiration is basically the oxidation of the glucose to carbon dioxide and water with the release of energy. It can conveniently be divided into two parts:

1. Cellular (internal or tissue) respiration: the biochemical processes which take place within living cells that release the energy from glucose.
2. Gaseous exchange (external respiration): the processes involved in obtaining the oxygen needed for respiration and removing the gaseous wastes such as carbon dioxide.

Cellular respiration is a complex metabolic process of over seventy reactions; these may be conveniently divided into three main stages:

1. Glycolysis
2. Krebs' (tricarboxylic acid) cycle
3. Electron transfer system.

Electron (hydrogen) Carriers

Before discussing these stages, it is necessary to mention a group of substances that are employed throughout cellular respiration. These are the electron (hydrogen) carrier or electron (hydrogen) acceptor molecules. Many reactions in cellular respiration involve the release of electrons at a high energy level.

These electrons are collected by electron carrier molecules which pass them to electron carriers at lower energy levels. The energy released is used to form ATP from ADP. The electrons are initially released as part of a hydrogen atom which later splits into a proton and an electron. This is why the carriers have the alternative name hydrogen carriers.

They act as coenzymes because they are essential to the functioning of the dehydrogenase enzymes that catalyse the removal of the hydrogen atoms. One of the most important is nicotinamide adenine dinucleotide (NAD) and others include nicotinamide adenine dinucleotide phosphate (NADP), flavoprotein (FP) and cytochromes.

Glycolysis

The first stage in the breakdown of glucose, glycolysis, is common to all organisms. In anaerobic organisms it is the only stage in respiration. It occurs in the cytoplasm of the cell. At this level it is not necessary to know all the intermediate compounds or enzymes but an overall appreciation of the major features is required.

The process of glycolysis produces two molecules of pyruvic acid for each molecule of glucose degraded. Effectively every substance after Stage 4 is doubled in quantity for each glucose molecule.

The total energy yield is therefore two molecules of ATP directly and six molecules of ATP produced from the two reduced NAD molecules. A total of eight ATP molecules.

Krebs' (tricarboxylic acid) cycle

The pyruvic acid formed as a result of glycolysis may in the absence of oxygen be converted to a variety of substances to yield a little energy.

These processes constitute the anaerobic pathways. In the presence of oxygen the pyruvic acid enters the Krebs' cycle. Before entering the actual cycle one of the three carbon atoms of pyruvic acid is oxidized to carbon dioxide and a molecule of NAD is reduced by the addition of two hydrogen atoms (the NAD yields a further three ATPs).

This leaves an acetyl group (CH_3CO) which is readily accepted by a coenzyme called coenzyme A. The resulting substance is acetyl coenzyme A. The 2-carbon acetyl group of this compound combines with a 4-carbon substance called oxaloacetic acid to give the 6-carbon molecule citric acid.

In a series of reactions two carbon dioxide molecules are produced and the 4-carbon oxaloacetic acid molecule is regenerated in readiness to receive another 2-carbon acetyl group from acetyl coenzyme A. Other products include a total of eight hydrogen atoms which are used to reduce three molecules of NAD and one molecule of FP.

Stages of glycolysis

Stage 1

The glucose molecule has a phosphate group added to make it more reactive (= activation). The phosphate group is donated by ATP

Stage 2

The glucose phosphate is reorganized into a fructose phosphate molecule

Stage 3

The fructose phosphate is further activated by the donation of a second phosphate group by an ATP molecule

Stage 4

The 6-carbon fructose diphosphate is split into two 3carbon triose phosphate molecules

Stage 5

Hydrogen atoms are removed from the triose phosphate molecules and taken up by NAD. Inorganic phosphate is added to further activate the triose phosphate

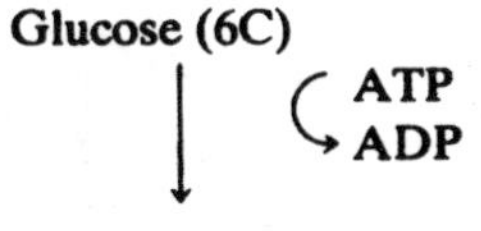

Glucose phosphate (6C)

↓

Fructose phosphate (6C)

↓ (ATP → ADP)

Fructose diphosphate (6C)

Triose phosphate (3C) ← **Triose phosphate (3C)**

↓ (**Inorganic phosphate**; **2 NAD (oxidized)** → **2 NAD (reduced)**)

2 × Triose diphosphate (3C)

↓ (**2 ADP** → **2 ATP**)

2 × Triose phosphate (3C)

↓ (**$2H_2O$**; **2 ADP** → **2 ATP**)

2 × Pyruvic acid (3C)

Stage 6

A phosphate molecule is lost and ATP is regenerated

Stage 7

A phosphate molecule is lost and ATP is formed; a water molecule is also lost, for each triose phosphate

These reduced electron (hydrogen) carriers eventually pass on the hydrogen atoms to oxygen yielding 11 more ATP molecules for each pyruvic acid. In addition a further ATP molecule is yielded directly during the cycle to give a total of 12 ATPs per pyruvic acid molecule.

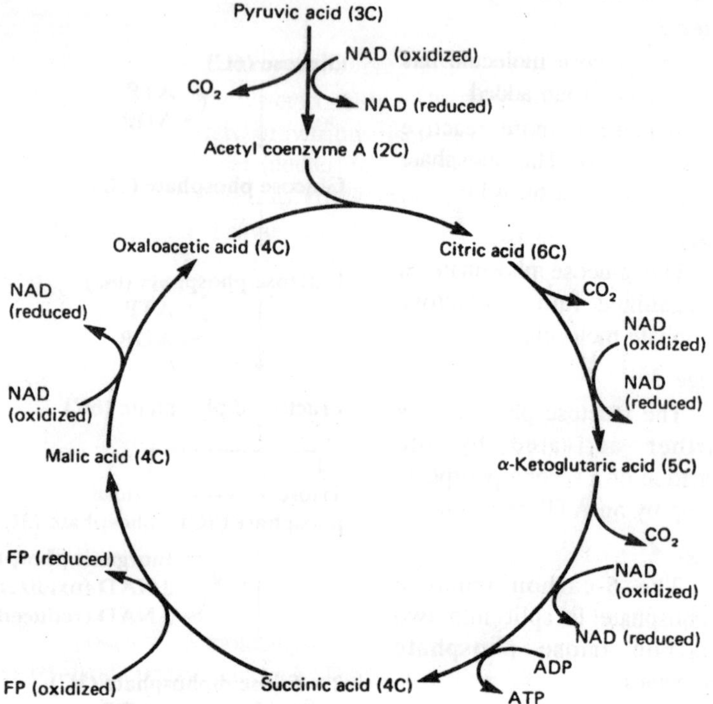

Figure 8.1: Summary of Krebs' cycle.

Differential centrifugation may be used to separate parts of a cell. By testing each part it can be shown that the Krebs' cycle takes place in the mitochondria.

Importance of Krebs' cycle

1. It provides hydrogen atoms which ultimately yield the major part of the energy derived from the oxidation of a glucose molecule.
2. It is a valuable source of intermediates which are used to manufacture other substances, e.g. fatty acids, amino acids, carotenoids.

Electron Transfer System

Although the glucose molecule has been completely oxidized by the end of the Krebs' cycle, much of the energy is in the form of hydrogen atoms which are attached to the hydrogen carriers NAD and FP.

These atoms are passed along a series of carriers at progressively

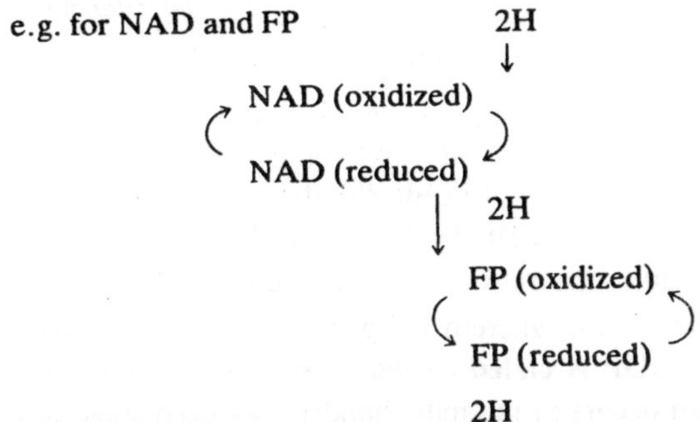

The full pathway can be summarised.

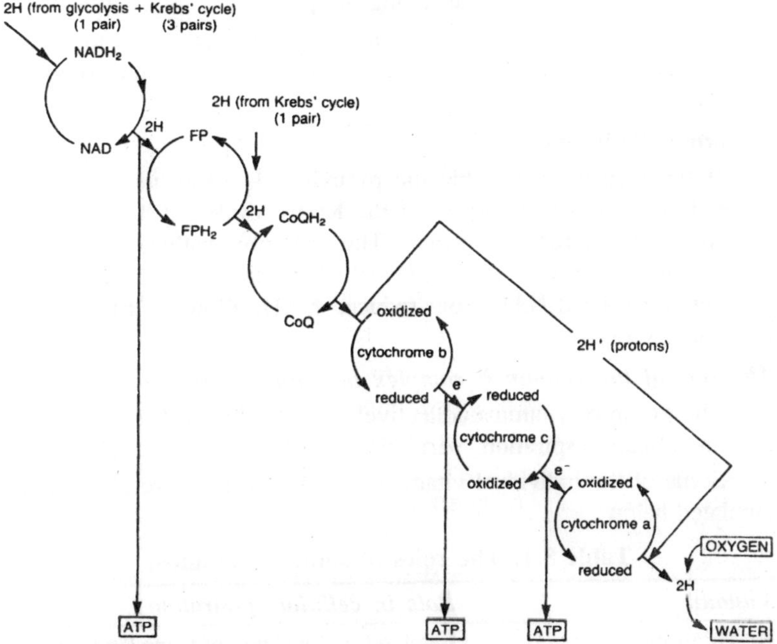

Figure 8.2: The electron transfer system.

lower energy levels. As they lose their energy it is harnessed to produce ATP molecules, three for each molecule of NAD and two for each one of FP. The other carriers in the system are iron-containing proteins called cytochromes.

The hydrogen atoms split into their protons and electrons during the pathway and the electrons are carried by the cytochromes. They recombine with their protons before the final stage where the newly reformed hydrogen atoms combine with oxygen to form water.

Although oxygen, so closely connected with aerobic respiration, only plays a role at this final stage, it is nevertheless vital since it drives the whole process. Except for the ability of oxygen to act as the final hydrogen acceptor the hydrogens would accumulate and the process of aerobic respiration would cease.

The whole process whereby oxygen effectively allows the production of ATP from ADP is called *oxidative phosphorylation*. The electron transfer system occurs in the mitochondria. At each stage of the system the hydrogen atoms (or electrons) are passed from the reduced carrier to the oxidized carrier further along the pathway.

Thus the reduced carrier becomes oxidized and able to accept more hydrogen atoms (or electrons) and the oxidized one becomes reduced.

Anaerobic Pathways

If no oxygen is available the pyruvic acid molecules formed at the end of glycolysis do not enter the Krebs' cycle but follow one of a number of anaerobic pathways. The anaerobic pathways are often referred to as fermentation. The pathways often use up 2H and so regenerate oxidized NAD from reduced NAD, allowing it to be used again in glycolysis.

The role of the vitamin B complex in cellular respiration

The group of vitamins collectively called vitamin B play a major role in cellular respiration, particularly by acting as coenzymes.

Some of the individual vitamins and their roles in respiration are tabulated below:

Table 8.1: The roles of some B vitamins.

Vitamin	*Role in cellular respiration*
B_1 Thiamine	Involved in formation of some Krebs' cycle enzymes. Forms part of acetyl coenzyme A
B_2 Riboflavin	Forms part of the hydrogen carrier flavoprotein (FP)
B_3 Niacin (nicotinic acid)	Forms part of the coenzymes NAD and NADP. Forms part of acetyl coenzyme A
B_5 Pantothenic acid	Forms part of acetyl coenzyme A

ATP and energy yields

Adenosine triphosphate (ATP) is the form in which energy from the breakdown of glucose is temporarily stored. ATP consists of the organic base adenine, the 5-carbon sugar ribose and three phosphate groups.

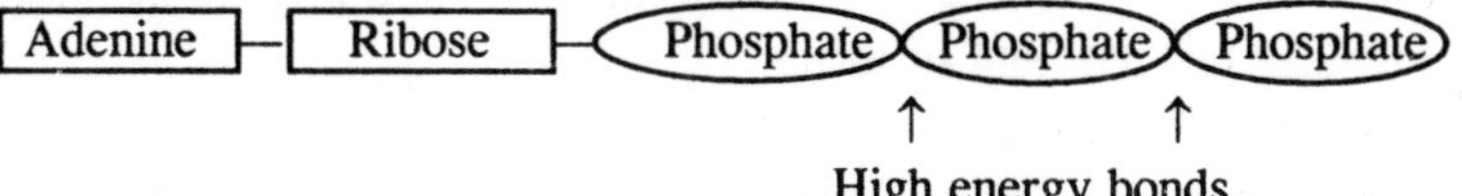

The removal of the final phosphate to form adenosine diphosphate (ADP) releases about 34 kJ/mole of energy.

$$\text{ATP} + H_2O \rightarrow \text{ADP} + \text{phosphate} + 34 \text{ kJ}$$

Although the terminal bonds are referred to as 'high energy bonds' the term is misleading because the energy is not within these bonds but spread throughout the molecule.

In fact, energy is only released because there is a net lowering of energy when one bond is broken and another reformed. High energy bonds as such do not exist; they are a simplified way of expressing a more complex process.

The amount of ATP produced during anaerobic respiration is small. For instance in alcoholic fermentation glycolysis yields two molecules of ATP directly and two hydrogen atoms. However the hydrogen atoms do not enter the electron transfer system as they are used to form the alcohol.

The total yield is therefore only two ATPs or 68 kJ of energy from a potential of nearly 3000 kJ for the complete oxidation of a glucose molecule. The process is therefore only about 2% efficient.

In aerobic respiration the yield per glucose molecule is:

Glycolysis	=	8 ATP (2 ATP molecules directly + 2 molecules of reduced NAD which yields a further 6 ATP)
Pyruvic acid→Acetyl CoA	=	6 ATP (2 molecules of reduced NAD each yielding 3 ATP)
Krebs' cycle	=	24 ATP (6 molecules of reduced NAD each yielding 3 ATP +2 molecules of reduced FP yielding 2 ATP + 2 molecules of ATP formed directly)
Total ATP yield	=	38 molecules of ATP.

This gives a total energy yield of 38 × 30.6 kJ = 1162.8 kJ, which is about 40.3% efficient.

One problem arising from the totalling of these ATPs that causes confusion is that the yields refer to the number of ATPs from a single glucose molecule. Early in glycolysis the 6-carbon sugar is split into two 3-carbon sugars. The yields from this point onwards are therefore doubled to take account of this.

Respiratory Quotients

A respiratory quotient (RQ) is a measure of the ratio of carbon dioxide evolved to the oxygen consumed.

$$RQ = \frac{CO_2 \text{ evolved}}{O_2 \text{ consumed}}$$

For a hexose sugar such as glucose it can be seen from the equation

$$C_6H_{12}O_6 + 6O_2 \rightarrow 6CO_2 + 6H_2O$$

that the ratio is $\frac{6CO_2}{6O_2} = 1.0$.

For a fat such as stearic acid however the equation for its complete oxidation is

$$C_{18}H_{36}O_2 + 26O_2 \rightarrow 18CO_2 + 18H_2O$$

$$\text{The RQ} = \frac{18CO_2}{26O_2} = 0.7$$

Other RQs are

Malic Acid—1.33

Proteins—0.9 (Though this varies slightly according to the particular protein)

In practice organisms rarely oxidize one compound alone and therefore RQs may lie between these values, e.g. for humans the RQ is typically 0.85 indicating the oxidation of a mixture of carbohydrate (1.0) and fat (0.7) rather than protein which is normally only respired in extreme circumstances, e.g. starvation.

GASEOUS EXCHANGE

Underlying Principles

Much of the energy utilized by organisms for osmotic, mechanical, electrical and chemical work is derived from ATP formed in the mitochondria during oxidative phosphorylation.

This process requires oxygen which the organism must obtain from its environment of air or water. Regardless of source, the oxygen enters the organism by diffusion.

In order to obtain it in adequate quantities and remove the carbon dioxide produced, a relatively large, moist surface must be exposed to a source of oxygen. The distance between this oxygen source and the cells requiring or transporting it must be small and the diffusion gradient must be maintained if the process is to satisfy the needs of the organism.

In small organisms such as protozoa and unicellular algae the surface area is sufficiently large compared with volume so that diffusion over the whole body surface satisfies their needs.

As organisms became multicellular and grew in size they were only able to utilize the whole body surface for obtaining oxygen if their energy demands were small and their shape became flattened (e.g. platyhelminthes) or the centre contained no respiring material (e.g. coelenterates).

Any further increase in size or metabolic rate required some specialized gaseous exchange system to compensate for a lower surface area/volume ratio and greater oxygen demand. Such systems included

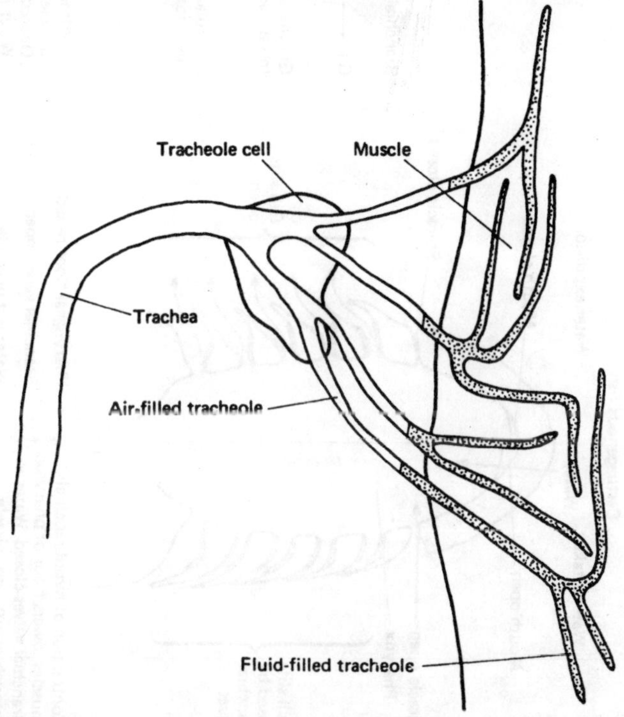

Figure 8.3: Tracheoles in an insect.

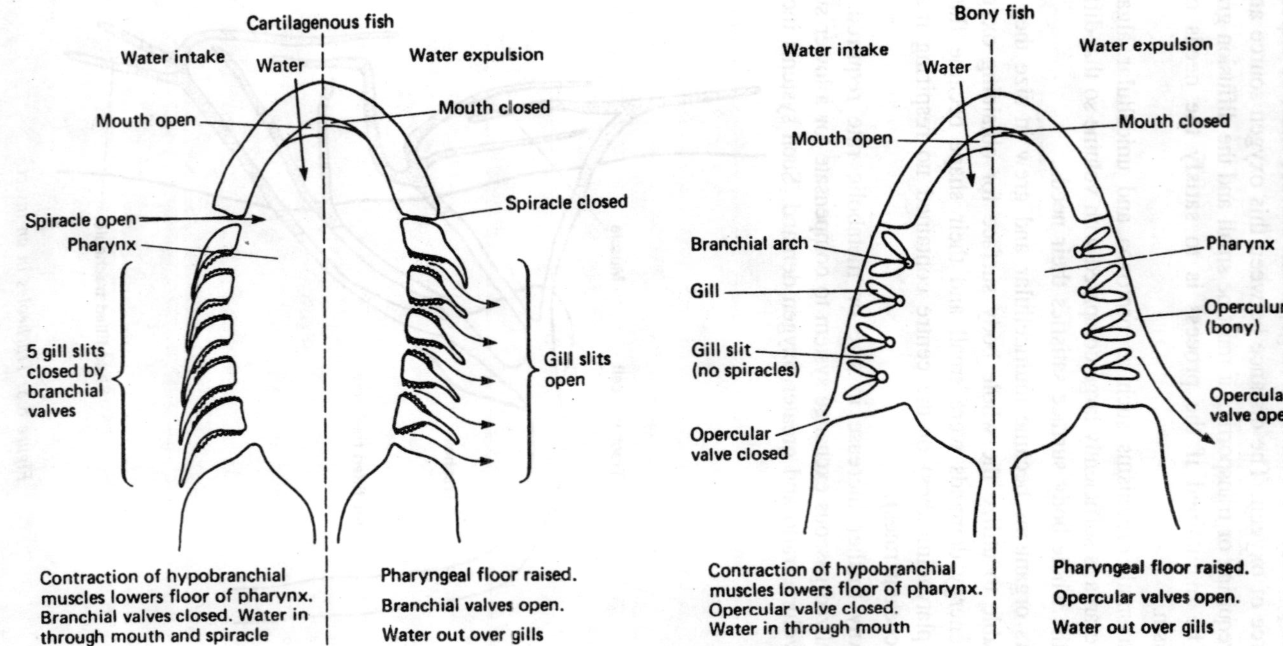

Figure 8.4: Tracheoles in an insect.

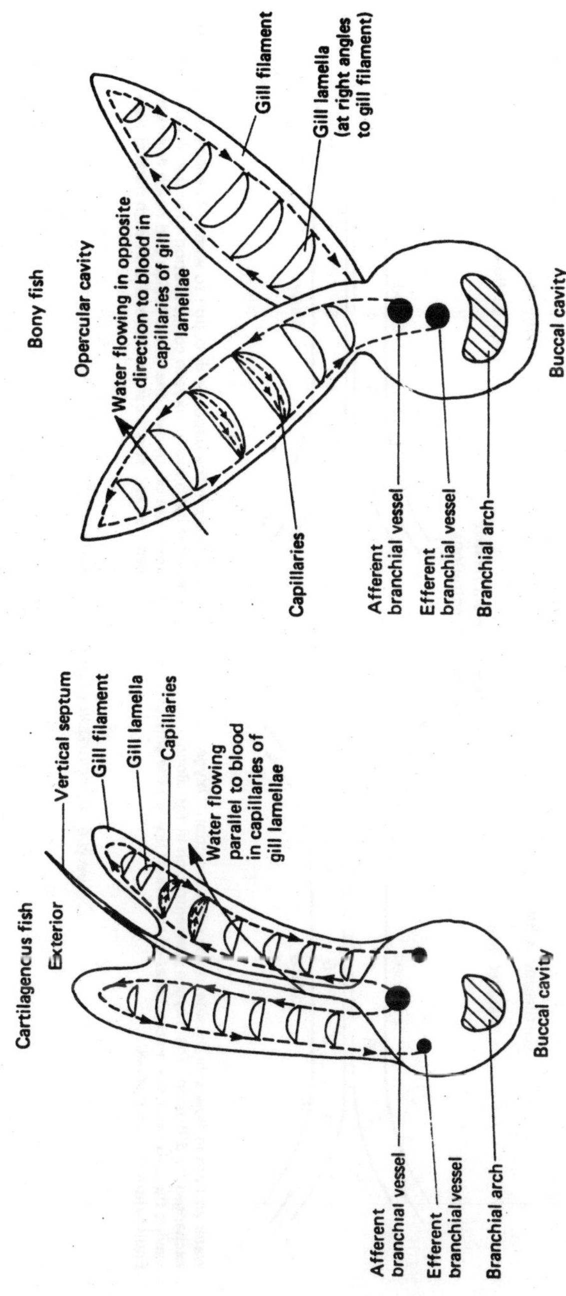

Figure 8.5: Tracheoles in an insect.

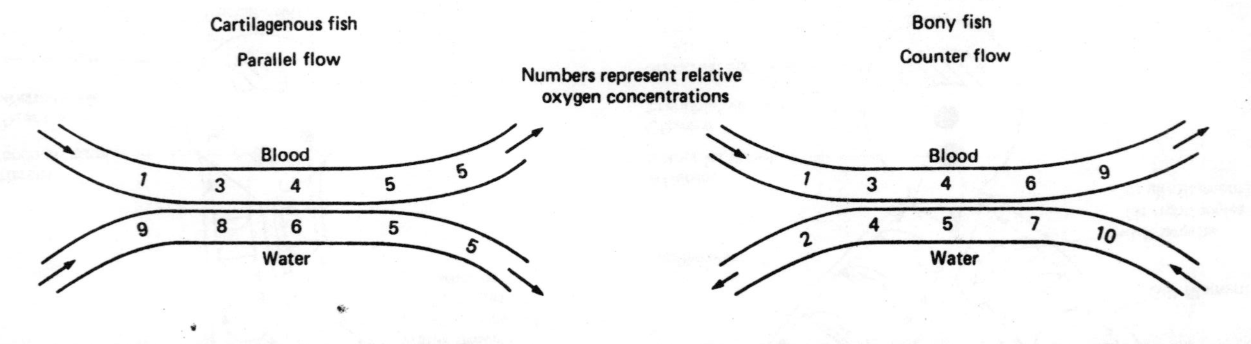

Figure 8.6: Comparison of parallel and counter flow systems.

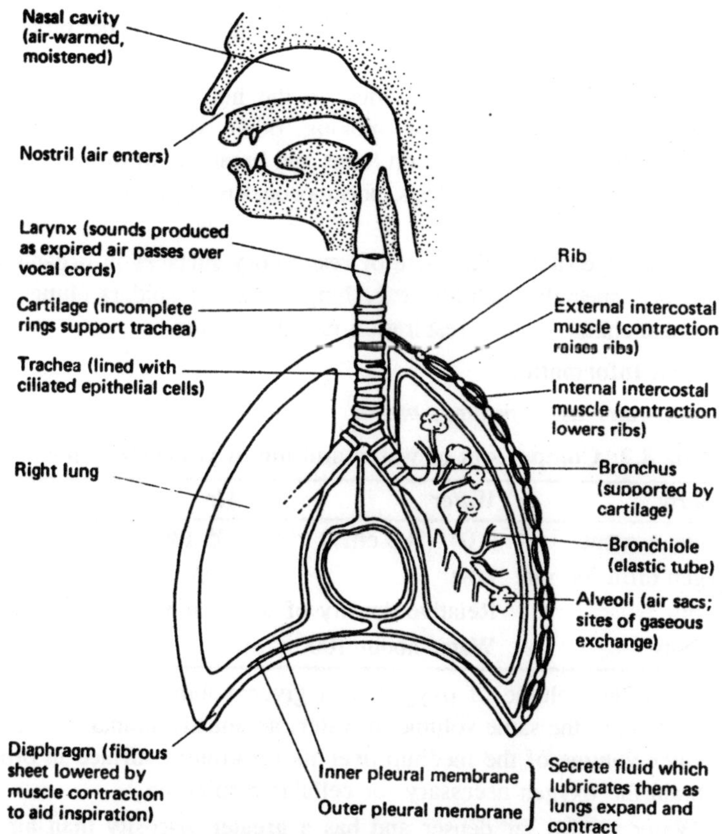

Figure 8.7: Human respiratory system.

tubular ingrowths to carry' air directly to the cells (e.g. insects) or organs of gaseous exchange situated in one part of the organism and supplied with a means of transporting gases from it, to and from the other cells.

In water this second system takes the form of gills and on land lungs are used. Whichever system is used the respiratory surface must have a constant supply of the oxygen-carrying medium flowing over it, to maintain the diffusion gradient.

Points of Perspective

It would be useful for candidates to have seen mammalian lungs and fish gills, either by dissection or demonstration. As this topic lends itself quite well to data type questions, it would be beneficial for candidates to have interpreted such data, or even to learn some figures, e.g. oxygen consumption of different animals and under different levels of activity.

The effects of certain activities on the human lungs would be useful to know, e.g. effects of smoking, other types of atmospheric pollution, as well as the effect on lungs and breathing of diseases such as cancer, bronchitis, tuberculosis, pneumoconiosis, silicosis, emphysema and asbestosis.

Knowledge of a wide range of respiratory surfaces in organisms would be especially helpful, e.g. lung books in spiders, lungs of pulmonate snails, cloacal respiration in holothurians.

Essential Information

Air and water as respiratory media

Table 8.2: Comparison of water and air as respiratory media.

Property	*Water*	*Air*
Oxygen content	0.04 - 9.0 $cm^3/1$	105-130 cm 3/1
Oxygen diffusion rate	Low	High
Density	Relative density of water about 1000 × air	
Viscosity	Water about 100 × air	

Since the volume of oxygen in a given volume of air is much greater than in the same volume of water, an aquatic animal must pass a greater volume of the medium over its respiratory surface in order to obtain the oxygen necessary for cellular respiration.

Water is also far denser and has a greater viscosity than air at the same temperature and so it is more difficult for the organism to extract the necessary oxygen from it.

It would, therefore, seem easier for animals to obtain oxygen from air than from water but this is not the case since terrestrial animals must avoid desiccation and therefore have relatively impermeable skins. Terrestrial animals and the more active aquatic ones do not rely on the general body surface for respiratory exchange but have developed specialized respiratory surfaces.

The simplest of these surfaces is gills which may be external, e.g. annelida, amphibian tadpoles, or internal, e.g. fish. Insects have a tracheal system. Terrestrial vertebrates and some invertebrates, e.g. pulmonate snails, have developed lungs. For maximum efficiency internal respiratory surfaces such as internal gills and lungs need to be ventilated.

General properties of respiratory surfaces

1. Large surface area/volume ratio: the body surface is adequate in very small organisms; infoldings of a restricted part of the body provide a large respiratory surface, e.g. lungs and gills, in larger organisms.
2. Moist: surfaces are moist so that diffusion may occur in solution.
3. Thin: diffusion is only efficient over short distances because the rate of diffusion is inversely proportional to the distance between the concentrations on the two sides of the respiratory surface.
4. Transport: this must occur in order to maintain the diffusion gradient. In large metazoans an efficient vascular system is required.

Mechanisms of Gaseous Exchange

Terrestrial flowering plants

By virtue of their autotrophic mode of nutrition the supply of gaseous oxygen to the tissues of flowering plants is facilitated for three reasons:

1. When photosynthesizing oxygen is produced as a by-product and may be used directly for respiration.
2. Photosynthesis requires specialized structures—the stomata—in the leaves to allow diffusion of carbon dioxide and these can be utilized to obtain oxygen from the atmosphere when no photosynthesis is taking place.
3. The roots have a thin surface and large surface area to facilitate the collection of water by osmosis for photosynthesis. This feature allows oxygen to diffuse into the root easily from the soil spaces.

The stems of plants obtain their oxygen either through stomata (herbaceous flowering plants) or lenticels (woody flowering plants). The central tissue of woody stems and roots is composed of dead xylem cells.

Therefore no living cell of a plant is ever far from a source of oxygen and no specialized respiratory surface is required as diffusion over the whole surface will suffice.

Insects-tracheal system

The respiratory system of an insect comprises a series of tubes, tracheae, with paired openings to the exterior called spiracles. The tracheae branch to form terminal tracheoles, often less than 1.0 μm in

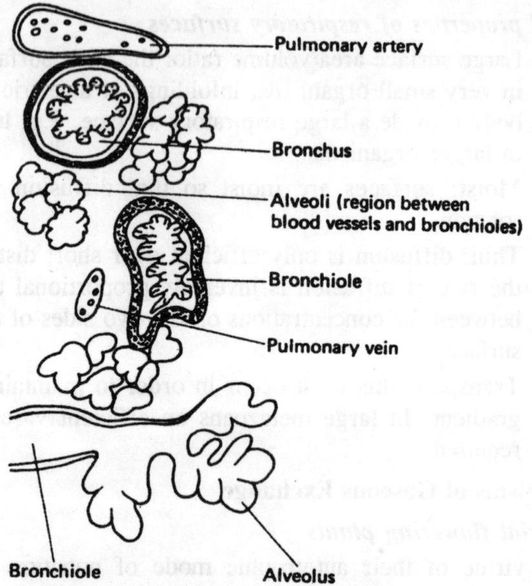

Figure 8.8: LS part of mammalian lung.

diameter which reach more or less every part of the insect's body. In insects blood does not carry oxygen and carbon dioxide.

Except at their very ends the tracheae and tracheoles are filled with air; oxygen and carbon dioxide are transported primarily by diffusion. Ventilation of the larger tracheae occurs through movements of the muscles or exoskeleton.

The spiracles are surrounded by hairs to prevent the entry of dust and parasites and they may be opened and closed by valves. The larger tracheae are lined by cuticle which must be shed at each moult and this, together with the rate of diffusion in narrow tubules, is the main factor limiting the size of insects.

During exercise lactic acid accumulates in the muscle, raising its osmotic pressure. Water passes from the tracheole into the muscle by osmosis thus drawing air further down the tracheole.

Fish - gills

Although fundamentally similar, there are some differences in structure, and therefore in mechanisms of gaseous exchange, between cartilagenous and bony fish.

Mammal (e.g. man) -lungs

Ventilation, which is clearly essential for a respiratory surface so

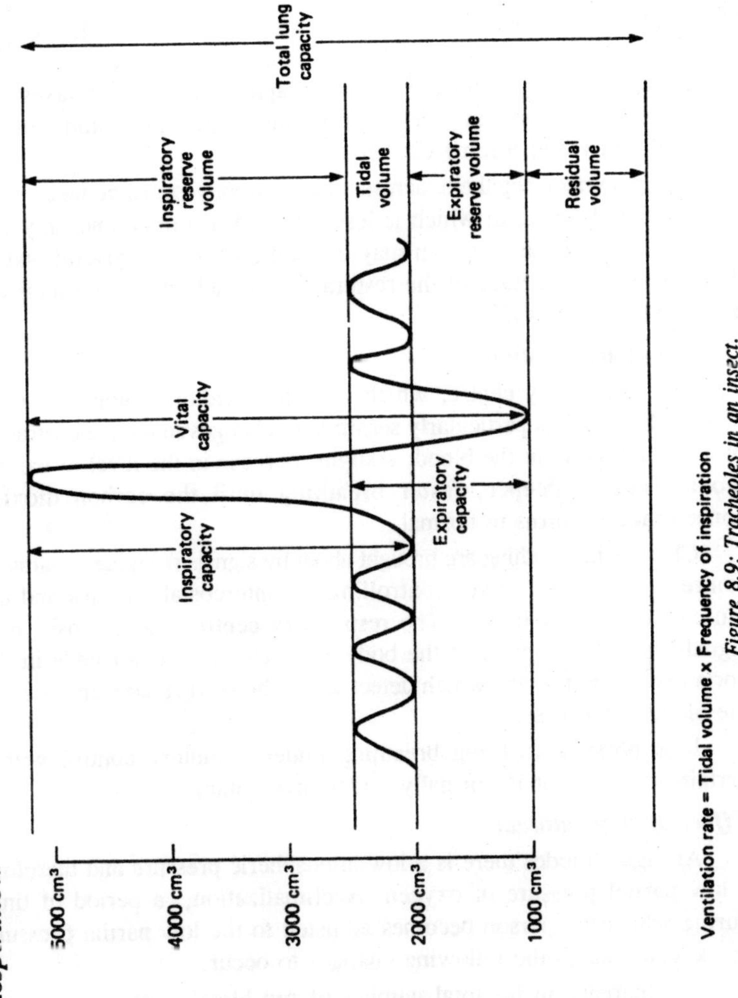

Figure 8.9: Tracheoles in an insect.

deep inside the body, is brought about by the diaphragm and the ribs. The diaphragm is a fibrous sheet of tissue whose muscular edges are attached to the thoracic wall.

Contraction of these muscles flattens the diaphragm and increases the volume of the thoracic cavity. Between the ribs are the intercostal muscles whose contractions cause the ribs to move. The external intercostal muscles contract to move the ribs upwards and outwards, at the same time raising the sternum.

This also causes the volume of the thoracic cavity to increase. As the volume increases the pressure decreases to below that of the atmosphere and air rushes in. This is inspiration (breathing in). Expiration is brought about when the diaphragm muscle relaxes and the internal intercostal muscles contract decreasing the volume of the thoracic cavity and forcing air out.

The respiratory epithelia across which gaseous exchange takes place are the alveoli, each of which is less than 0.5 μm thick and only 100 μm across. Each lung of man may contain 350 million alveoli and so the total internal surface of the respiratory epithelium is enormous (an estimated 80-100 m^2).

Control of Respiration

The respiratory centre, which is in the medulla oblongata at the base of the brain, is particularly sensitive to changes in the concentration of carbon dioxide in the blood; a slight increase in the level of carbon dioxide causes deeper, faster breathing until the carbon dioxide concentration returns to normal.

Changes in breathing are brought about by signals from the respiratory centre to the spinal nerves controlling the intercostal muscles and the muscles of the diaphragm. The respiratory centre is also sensitive to signals from other parts of the body, e.g. chemoreceptor cells in the aorta and carotid artery which detect when the oxygen concentration in the blood decreases.

It is possible to bring breathing under voluntary control within certain limits but it is normally under involuntary control.

Effect of high altitude

At high altitudes there is a low atmospheric pressure and therefore a low partial pressure of oxygen. Acclimatization,, a period of time during which the person becomes adjusted to the low partial pressure of oxygen, causes the following changes to occur:

1. Increase in the total number of red blood cells
2. Increase in the total haemoglobin content
3. The ventilation rate increases
4. Cardiac frequency increases.

Diving mammals

Relative to their size no diving mammals have lungs significantly larger than those of man but they do have a greater blood volume, e.g. in man blood is about 7% of the body weight; in diving marine mammals it is about 10-15% of the body weight. Other differences include:

1. Enlarged blood vessels to act as reservoirs of oxygenated blood
2. Higher proportion of red blood cells
3. Myoglobin more concentrated
4. Slower heart beat (to conserve use of oxygen)
5. Reduction of blood supply to organs and tissues tolerant to oxygen deficiency, e.g. digestive system, muscles
6. Compression of air spaces, e.g. lungs, middle ehr, which protects the animal from the 'bends'
7. Respiratory centres do not function automatically to cause breathing at a certain concentration of carbon dioxide.

9

INTEGRATION

HORMONES AND HOMEOSTASIS

Underlying Principles

An organism needs methods of communication to co-ordinate its organ systems and behaviour. In an animal, information about both the environment and the body's internal state is integrated by the endocrine and nervous systems.

Appropriate responses to changes in the environment are essential for the animal's survival. Efficiency of organ systems in all animals is improved by the maintenance of a constant internal environment-homeostasis-so that metabolic processes can be regulated more easily.

All animals and plants have some ability to control their osmotic pressure, chemical constitution and body temperature, but homeostatic mechanisms are best developed in the birds and mammals.

Points of Perspective

Even if the syllabus does not mention it specifically, it would be useful for the candidate to be able to recognize microscope slides of the thyroid, pancreas and liver.

The presumed mechanism of hormonal action in the cell, via cyclic AMP, should be considered.

A knowledge of the effects, symptoms and treatment of excess and deficiency of all the hormones studied would be helpful background information, together with some idea of the experimental work which has led to the present understanding of endocrine function.

Information about the biochemical processes taking place in the liver, especially glycolysis and deamination, would also be useful.

Essential Information

Hormones and the endocrine system

Hormones are organic substances produced in one part of an organism and transported to another, where they exert their effects. In animals they are usually formed in the endocrine or ductless glands and secreted directly into the bloodstream, which carries them to their target organs.

The glands are integrated into the endocrine system, which interacts with the nervous system to provide the communication, co-ordination and control within the body.

Comparisons between the endocrine and nervous systems

At some point in both these systems a stimulus triggers the chemical transmission of a message; this produces a response. They each carry out these basic functions rather differently.

Table 9.1: Comparison of the endocrine and nervous systems.

	Hormonal communication	*Nervous communication*
Origin of stimulus	Gland	Sense receptor
Nature of stimulus	Hormone	Nervous impulse
Means of transmission	Bloodstream	Nerve fibre
Destination of stimulus	All over body	To a specific point
Receptor	Target organ	Effector (muscle or gland)
Speed of transmission	Usually slow	Rapid
Effects	May be widespread	Localized
Duration	Usually long-lasting	Usually brief

Hormones and Glands

The pituitary gland at the base of the brain co-ordinates the activities of most of the other endocrine glands, and is itself a unit functionally integrated with the hypothalamus.

The hypothalamus constantly monitors the composition of the blood and in response to any changes produces factors which affect hormonal release from the pituitary.

Specialized cells in the gut produce the digestive hormones gastrin, secretin and cholecystokininpancreozymin, which stimulate the release of digestive enzymes.

Feedback Mechanisms

These are control mechanisms in which the response to a stimulus

Table 9.2 Some hormones and their effects

Gland	*Hormone*	*Effects*
Pituitary Anterior lobe	Somatotrophin (Growth hormone, GH)	Increases growth rate in young animal and maintains size of body parts in adult
	Thyroid stimulating hormone (TSH)	Acts on thyroid gland, thereby controlling metabolic rate
	Adrenocorticotrophic hormone (ACTH)	Controls activity of adrenal cortex
	Melanocyte stimulating hormone (MSH)	Darkens skin in frogs; function in humans unknown
	Follicle stimulating hormone (FSH)	Together known as the gonadotrophins, control development, maintenance and hormonal functions of the gonads.
	Luteinizing hormone (LH), or interstitial cell stimulating hormone (ICSH)	
	Prolactin, or luteotrophic hormone (LTH)	Mammary gland development and milk production; maternal behaviour in birds
Posterior lobe	Anti-diuretic hormone	Stimulates water reabsorption from the kidney tubules (ADH)
	Oxytocin	Causes contraction of smooth muscles in uterus of a pregnant female

(Table 9.2 contd.)

Gland	Hormone	Effects
		and the release of milk during suckling
Thyroid	Thyroxine	Increases metabolic rate; induces metamorphosis in frogs
	Calcitonin	Lowers plasma calcium level
Parathyroids	Parathormone	Raises plasma calcium level
Pancreas	Insulin	Lowers blood glucose level
	Glucagon	Raises blood glucose level
Adrenals	Aldosterone	Stimulates sodium reabsorption from the kidney tubules Cortex
	Cortisol	Helps body resist stress, partly by raising blood glucose level
Medulla	Adrenaline	Prepare body for activity
	Noradrenaline	
Ovaries Testes		

itself affects that stimulus. In *negative feedback*, the response reduces the strength of the original stimulus, while in *positive feedback* the opposite occurs and the response increases the stimulus.

Negative Feedback

This type of feedback is very important in the maintenance of homeostasis. Its self-regulating characteristics make it an especially useful method for controlling release of hormones.

(a) Negative feedback

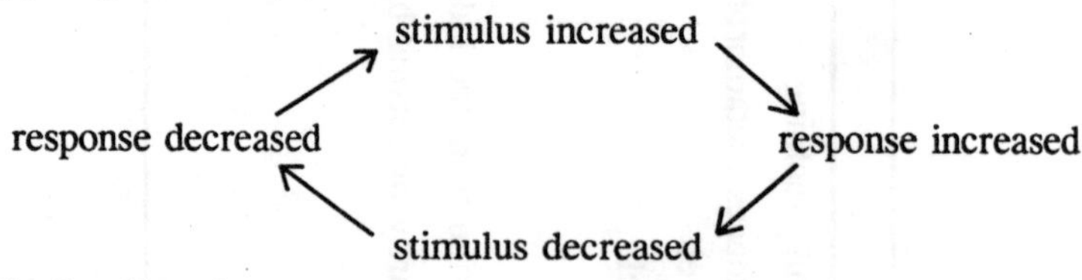

(b) Insulin release

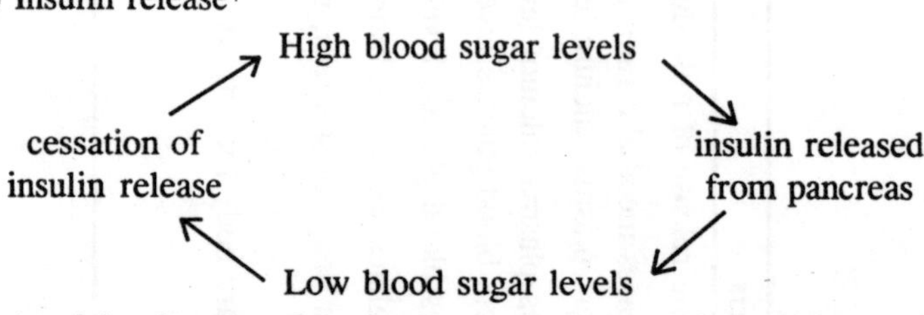

Lack of insulin therefore leads to permanent high blood sugar levels. As insulin reduces blood sugar partly by stimulating the active transport mechanism of glucose entry into the cell, the cells are deprived of glucose and diabetes results.

(c) Anti-diuretic hormone release

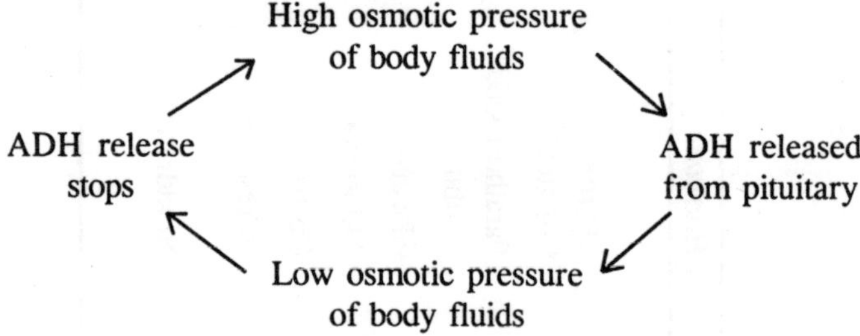

Osmoreceptor cells in the hypothalamus monitor water content of the body fluids and therefore control ADH release.

Positive feedback

As a control method, this is less often found, as it intensifies both stimulus and response. However it is put to good use as the smooth muscle of the uterus contracts during labour. Oxytocin stimulates contraction, and contraction, in turn, stimulates the release of more oxytocin. Consequently, contractions become more and more violent.

The liver

Many homeostatic functions are performed by the liver. The most important ones include:

1. *Maintenance* of steady blood sugar levels by the conversion of glucose to glycogen and *vice versa,* under the influence of insulin and glucagon
2. *Breakdown* or removal of excess or used materials such as
   (a) old red blood cells by *phagocytosis*
   (b) excess *lipids*
   (c) excess *cholesterol*
   (d) sex hormones and adrenaline
   (e) excess amino acids by deamination
   (f) poisons by detoxification
3. *Manufacture* of
   (a) plasma proteins such as fibrinogen
   (b) cholesterol
   (c) bile
4. *Storage* of iron, copper, potassium and vitamins A, D and B_{12}
5. *Production* of heat.

To be able to carry out these activities, the liver has a very rich blood supply. As well as oxygenated blood from the aorta it receives blood containing digested food in the hepatic portal vein, coming directly from the small intestine.

Temperature control in endotherms

Birds and mammals are capable of maintaining a relatively constant body temperature so that their metabolic rate need not vary with environmental temperature changes. This has been assisted by the evolution of feathers and fur for insulation. Mammals *raise* their body temperature by:

1. shivering
2. constriction of superficial blood vessels
3. raising their hairs to trap an insulating layer of air next to the skin surface
4. increasing metabolic rate

and they *lower* it by:

1. sweating and panting
2. dilation of superficial blood vessels
3. lowering hairs
4. decreasing metabolic rate

Blood temperature is monitored by the thermo-regulatory centre in the hypothalamus. Thermoreceptors in the skin signal changes in the environmental temperature via the sensory nerves to the brain.

Temperature control in ectotherms

All ectothermic animals rely on behavioural adaptations to avoid extremes of temperature. These adaptations are most marked in desert reptiles who are active in the early morning and evening, spend the hottest parts of the day in the shade and at night hide in crevices warmed by their own metabolic heat.

Temperature control in plants

Plants are more tolerant of temperature change than animals, but even so there is always an optimum temperature at which each species is most likely to thrive. As their metabolic rate is so low they are unable to raise their temperature much above that of the environment in cold weather, but in warm conditions transpiration may exert some cooling effect.

Adaptations to seasonal change

Animals and plants, especially those living in temperate parts of the world, must be able to adapt to seasonal changes if they are to survive. When conditions are good, usually in spring and summer, they can feed, grow, reproduce and disperse.

However in autumn a strategy must be adopted to see organisms through the winter when food may be unavailable. A common response is to become dormant. Many plants lose their leaves, while annuals overwinter as seeds.

Insects may survive in the egg or pupa stage. Small mammals, on account of their large surface area relative to volume, lose so much heat that it is difficult to find enough food to supply their needs; they are then obliged to hibernate, and their body temperature drops. Birds often migrate to areas where they can find food.

NERVOUS SYSTEM AND BEHAVIOUR

Underlying Principles

The nervous system permits rapid passage of information from one part of the body to another, so that suitable responses to stimuli can be made at once. In its simplest form it comprises a network of nerves connecting each part to every other part.

With an increase in the number of parts to be connected this system proved inadequate and a central nervous system (CNS) developed. Each part was connected to this central 'switchboard' which then connected incoming messages to the appropriate effector organs.

In bilaterally symmetrical animals it was appropriate to have the CNS running along the length of the body with paired nerves branching from it to each segment. As animals developed a definite head region which encountered stimuli before other parts, it obviously became the main location for sense organs.

To deal with this increased volume of nervous information at the anterior end, the CNS in this region swelled. The swelling or cerebral ganglion later developed into a brain whose function evolved beyond simple co-ordination to include storage of learned information and development of intelligent behaviour.

Points of Perspective

It would be an advantage to be aware of the experimental techniques which have been used in the study of brain function.

It is essential to be able to quote specific examples of reflex actions, and also to know the functions of the lobes of the cerebral

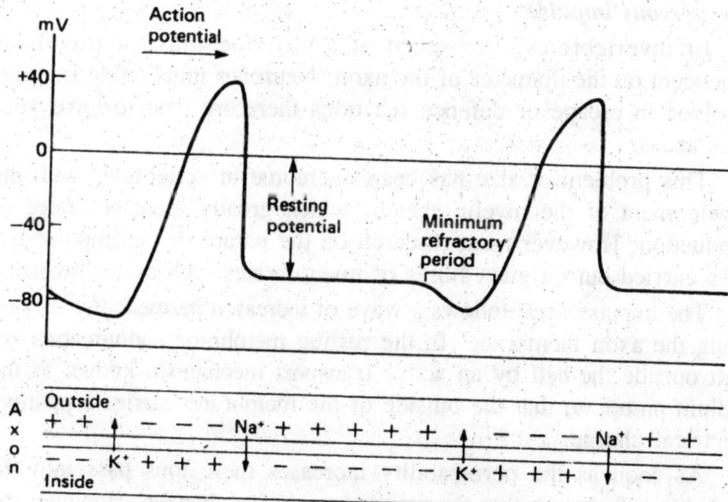

Figure 9.1: Transmission of action potential.

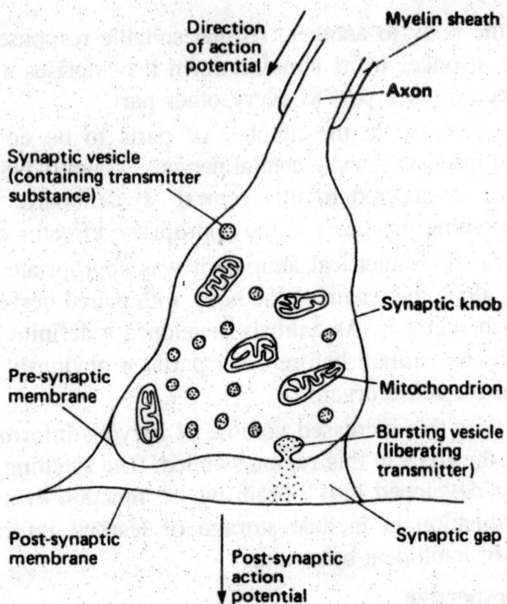

Figure 9.2: The synapse.

hemispheres. The differences between innate and learned behaviour patterns should be appreciated.

Essential Information

The nervous impulse

In invertebrates, the speed at which the impulse travels is dependent on the diameter of the axon. Neurones innervating muscles involved in escape or defence reactions therefore tend to have very wide axons.

This problem of size has been overcome in vertebrates with the development of the myelin sheath, which greatly increases rates of conduction. However most research on the nature of the impulse has been carried out on giant fibres of invertebrates such as earthworms.

The impulse itself follows a wave of increased permeability passing along the axon membrane. In the resting membrane sodium ions are kept outside the cell by an active transport mechanism known as the sodium pump, so that the outside of the membrane carries a positive electrical charge.

As soon as the permeability increases these ions pass into the axon by diffusion so that the membrane charge changes. However the membrane quickly recovers and the sodium ions are ejected. This

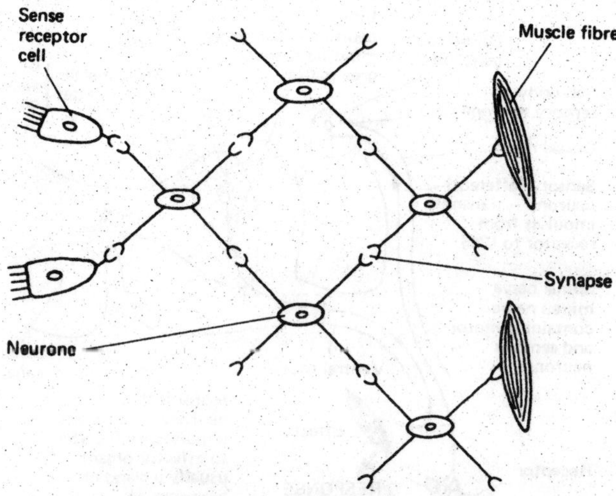

Figure 9.3: The coelenterate nerve net.

changing charge moving along the membrane is known as the *action potential*.

The Synapse

Neurones are not in direct contact with their neighbours; ti., re are tidy gaps known as synapses between them. The action potential stops at the synapse, but its arrival causes the release of a transmitter substance from the neurone which diffuses across the gap.

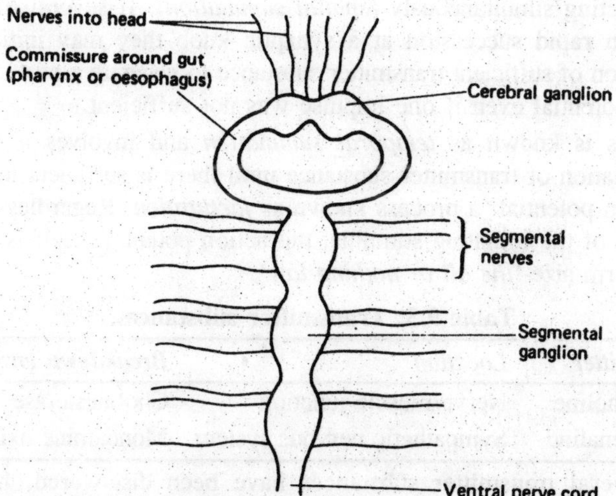

Figure 9.4: Nervous system of an annelid or arthropod.

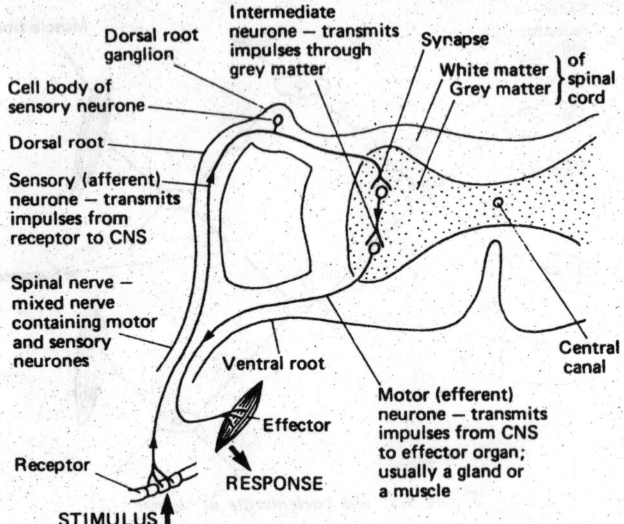

Figure 9.5: Reflex arc.

Arriving at the post-synaptic membrane, it causes a change in its permeability, so that a new action potential begins. The transmitter is then broken down enzymatically, to prevent it from continuously initiating action potentials.

Some stimuli do not initiate the production of enough transmitter substance to cross a synapse but enough may be produced by several knobs acting simultaneously (*spatial summation*). If several impulses arrive in rapid succession at a synaptic knob they may initiate the production of sufficient transmitter substance to generate a post-synaptic action potential even if one impulse was not sufficient.

This is known as *temporal summation* and involves a gradual accumulation of transmitter substance until there is sufficient to create an action potential, a process known as *facilitation*. Regardless of the strength of the initiating stimulus, the action potential itself is always of uniform size-'*the all or nothing law*'.

Table 9.3: Transmitter substances.

Transmitter	*Location*	*Breakdown enzyme*
Acetylcholine	Nerve/muscle junction	Cholinesterase
Noradrenaline	Sympathetic nervous system	Monoamine oxidase

Several transmitter substances have been discovered, but two important ones are acetylcholine and noradrenaline. Not all transmitters

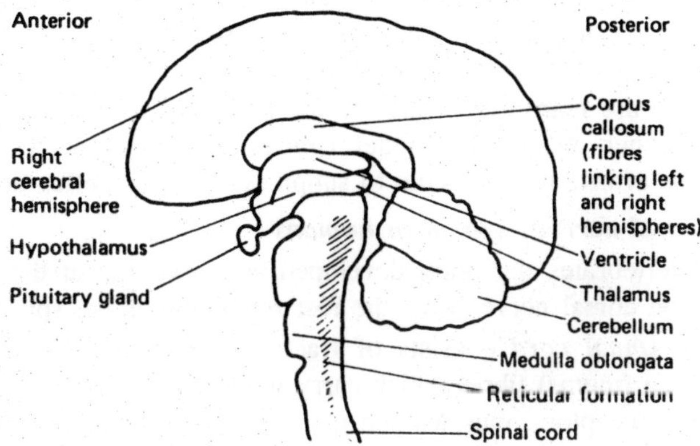

Figure 9.6: VS human brain.

excite the post synaptic membrane, some inhibit it so that it is more difficult for it to carry an impulse. Blocking of nervous pathways in this way helps the direction of impulses through the nervous system.

Invertebrate Nervous Systems

Coelenterates show all the elements of a nervous system necessary to perform reflex actions, i.e. receptors, neurones and effectors. They have multi-polar neurones forming a net in contact with both receptors and effectors. The arrival of one action potential at a synapse facilitates the passage of the next and so on until the whole net is stimulated.

This is known as *interneural facilitation*. In some coelenterates, e.g. sea anemones, there are also through conduction pathways which transmit a single impulse the length of the body. This provides the main function of a central nervous system: rapid transmission. *Annelids and arthropods* have a nervous system which is more centralized with a solid ventral nerve cord running the length of the body

This increases the speed with which impulses arriving from the receptors can be carried to the appropriate effectors. The nerve cord is swollen at the anterior end but this primitive brain plays little part in co-ordination, it merely deals with impulses arriving from the anterior receptors and messages to the anterior effectors.

Vertebrate Reflex Arc

A reflex is an automatic response to a stimulus, via the central nervous system.

RECEPTOR ↑ stimulus —sensory (afferent) neurone→ CNS —Motor (effector) neurone→ EFFECTOR ↑ response

In most reflex actions the sensory and motor neurones are connected via one or more intermediate neurones in th· brain or spinal cord (the knee jerk reflex is an exception to this).

There are many reflex arcs in the body and these are connected to each other and to the higher centres of the brain, so that, by conscious effort, it is possible to suppress a normal reflex action.

The central nervous system in humans

In vertebrates the brain developed as a swelling at the anterior end of the dorsal nerve cord, the rest remaining as the spinal cord.

The *spinal cord* consists of tracts of ascending (sensory) and descending (motor) fibres which carry information between the brain, and sense receptors, muscles and glands.

The *brain* may be subdivided into the following main regions each with their own specific functions:

Forebrain	olfactory tracts	concerned with sense of smell; found deep in forebrain
	cerebral hemispheres	very enlarged and important area of human brain. Electrical stimulation has led to 'mapping' of the most superficial area, the cerebral cortex
'Tweenbrain	thalamus	contains ascending and desce-nding tracts linking forebrain with spinal cord
	hypothalamus	largely controls the pituitary gland; seat of basic emotions or 'drives' such as hunger, thirst, fear, rage and sex.
Midbrain	corpora quadrigemina (optic lobes of lower vertebrates)	concerned with sense of sight
	red nucleus	helps control movement and posture
Hindbrain	medulla oblongata	controls voluntary movement and also has control centres for activities such as swallowing, salivation, heartbeat, vascular constriction and dilation, and respiration

cerebellum	regulation and co-ordination of muscular activities.

Stimulation of braincells

Neurones in the brain are normally stimulated by transmitter substances released at the synapse by a neighbouring cell. However chemical or physical changes in the body can sometimes affect brain cells directly. For example:

1. Osmoreceptor cells in the hypothalamus will stimulate the pituitary to release anti-diuretic hormone when the osmotic pressure of body fluids is low.
2. When carbon dioxide levels in the blood are high, reflex respiratory centres in the medulla increase breathing rate and depth.
3. Local temperature changes in the hypothalamus initiate the required responses to raise or lower body temperature.

The autonomic nervous system

This regulates the body's involuntary activities, and it is divided into the sympathetic and parasympathetic systems. The sympathetic system has comparable effects to adrenaline.

Table 9.4: Comparison of the sympathetic and parasympathetic systems.

Sympathetic system	*Parasympathetic system*
Prepares body for action	Prepares body for relaxation
Increases heartbeat	Slows heartbeat
Dilates arteries in skeletal muscles	Dilates arteries in gut
Slows gut movements	Speeds gut movements
Dilates bronchioles	Constricts bronchioles
Dilates pupil	Constricts pupil
Causes sweat glands to secrete	Causes tear and salivary glands to secrete Causes hairs to stand rect
Bladder and anal spincters contract	Bladder and anal spincters relax

Behaviour

It is difficult to categorize behaviour patterns rigidly but many of them involve elements of reflex action, orientation and learning.

Reflex action

Reflex actions have the advantage of being fast and automatic, and they are important in escape and avoidance reactions in all animals.

Orientation

Orientation often involves reflex actions but they are normally developed into behaviour patterns. There are two main forms of orientation found in animals:

Kinesis

This is an increase in random movement under unfavourable conditions. Woodlice move rapidly and turn frequently in dry conditions but the animals tend to congregate under humid conditions as their rate of movement slows.

Table 9.5: Comparison of learned and instinctive behaviour.

Instinctive behaviour	*Learned behaviour*
Inborn and not acquired during an animal's lifetime	Acquired during an animal's lifetime
Fixed and not adaptable although some minor modification over a modification over a long period may be possible	May be easily and rapidly adapted to suit changing circumstances
Similar amongst all members of a species	Varies considerably amongst different members of the same species
Unintelligent and there is no appreciation of the functions of the behaviour	May be intelligent and the animal often appreciates the function of a particular action
Often comprises a chain of actions in which the completion of one acts as the trigger for the start of the next	No fixed sequence of actions and the completion of one need not necessarily affect which action should follow
Apart from minor modifications the behaviour is permanent	Usually a temporary and short lived form of behaviour, although it may be reinforced, thus making it more or less permanent
Although there are many forms of instinct the basic form of this behaviour is the same for all organisms	Wide range of learned behaviour ranging from simple taxes or imprinting to complex forms of intelligence and reasoning

Taxis

This is the directional movement of the whole organism in direct response to a stimulus. Such stimuli may be light (phototaxis) or chemicals (chemotaxis), e.g. movement of planarians towards food; location of the ovum by the sperm of lower animals.

Learning

This is a change in behaviour based on experience and it is the most adaptable form of behaviour. Five subdivisions are often recognized:

Habituation

This is the diminishing of a response as a result of repeated stimulation. It is specific to a particular stimulus, e.g. snail, *Helix*, withdraws its tentacles in response to mechanical stimulation but ceases to do so after repeated stimulation.

Associative learning

In this the animal learns to associate a reward or punishment with a particular form of behaviour or stimulus. Such a pattern may be established either as a result of *classical conditioning* (the work of Pavlov on the salivation of dogs) or through *trial and error*.

Imprinting

Young animals tend to follow, or imprint on, their parents. Experiments have shown that they may follow the first moving thing they see.

Exploratory learning

Even without a specific reward or punishment an animal learns features of its environment which may benefit it later.

Insight learning

This involves the production of a new adaptive response as a result of 'insight'. It is a difficult subdivision to define and any definition of it results in controversy. It is thought to apply only to man.

Other factors affecting behaviour

Hormones may modify behaviour patterns; testosterone, for example, may increase the incidence of aggressive behaviour, and prolactin releases nestbuilding responses in birds.

Animals may show rhythmic changes in behaviour, according to the time of day or the season.

Characteristic behaviour patterns may be released when an animal meets another member of its own species. A male bird, meeting a female on his territory in spring, may begin a courtship display, but should he meet another male he will threaten him until the intruder withdraws, so that competition for food or a mate is removed.

Intraspecific interactions are especially important to humans, who owe their success largely to their ability to live in co-operative groups.

SENSE ORGANS

Underlying Principles

Animals have evolved specialized sense receptor cells to provide them with essential information about aspects of their environment. Touch, pressure, temperature and pain receptors are necessarily spread all over the body, while those for taste and smell are only in the mouth and nose.

Light, sound and gravity receptors are grouped together to form highly specialized sense organs.

Points of Perspective

An appreciation of the significance of colour vision to various animal groups, e.g. in food finding and courtship, is important.

An understanding of dark and light adaptation, and the formation of after images would be useful, together with colour blindness and defects of vision in humans.

Essential Information

Information about the environment is provided by receptors. Since there are many environmental stimuli the range of receptors must be varied. In man there are three main groups of receptors: chemoreceptors, photoreceptors and mechanoreceptors.

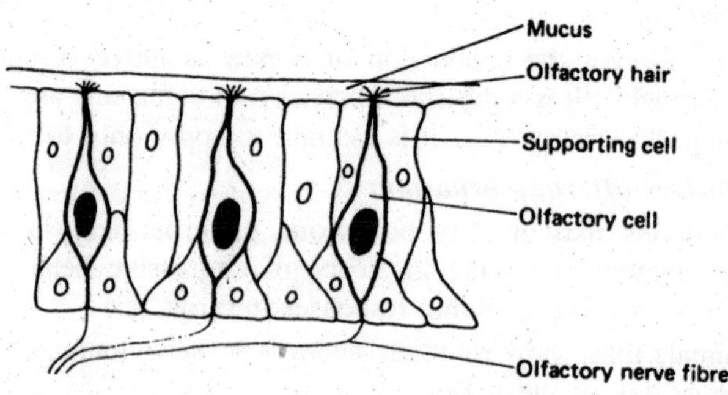

Figure 9.7: Olfactory epithelium.

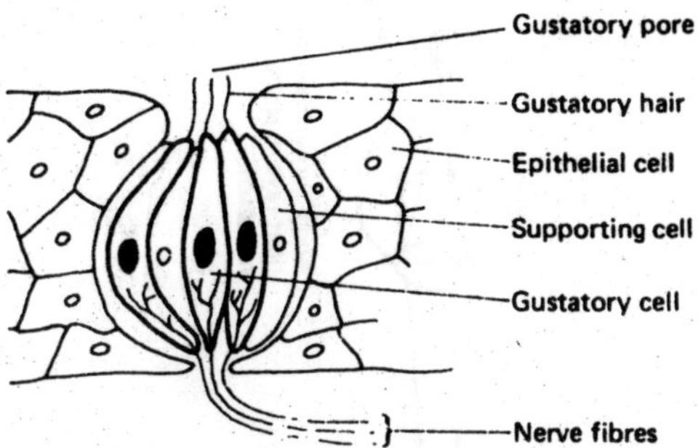

Figure 9.8: Taste cells.

Chemoreceptors

Olfactory epithelium

Olfactory cells are more or less evenly distributed between the supporting cells in two patches of epithelium in the upper nasal cavities. They are derived from the cell bodies of bipolar nerves. They respond to chemicals which are volatile at ordinary temperatures and which are soluble in fat solvents. The sense of smell is easily fatigued.

Taste cells

Taste cells are grouped into taste buds, found mainly round the base of the papillae of the tongue and also on the soft palate, epiglottis and around the opening of the oesophagus. Taste buds are sensitive only to chemicals which are soluble in water and although they appear similar each bud is sensitive to only one taste: sweet, sour, bitter or salt.

Photoreceptors

The eye As light enters the eye it is refracted by the curved, transparent cornea before passing through the pupil. This varies in size according to light intensity, as the muscles of the iris contract or relax:

Table 9.6: Action of the iris muscles.

Light	*Dim*	*Bright*
Circular muscles of iris	Relax	Contract
Radial muscles of iris	Contract	Relax
Pupil	Dilates	Constricts

The iris tends to be darkly pigmented in diurnal animals such as dogs, sparrows and cows, and pale in nocturnal ones such as cats and owls, so that more light can reach the retina.

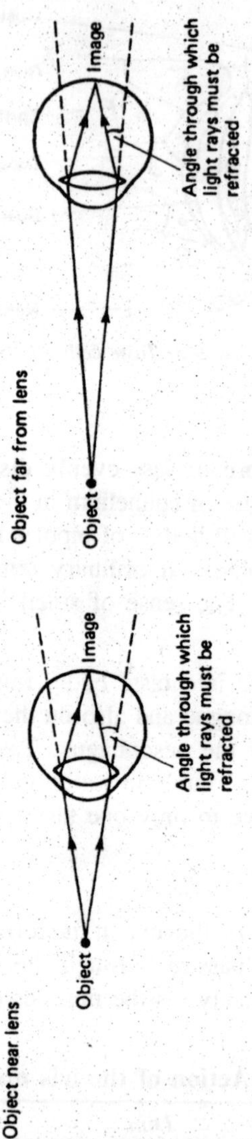

Figure 9.9: Accommodation.

Table 9.7: Accommodation of the eye

Object	*Near eye*	*Far from eye*
Light must be refracted	A lot	A little
Lens must be	Thick	Thin
Ciliary muscles	Contract	Relax
Suspensory ligaments	Slack	Tense

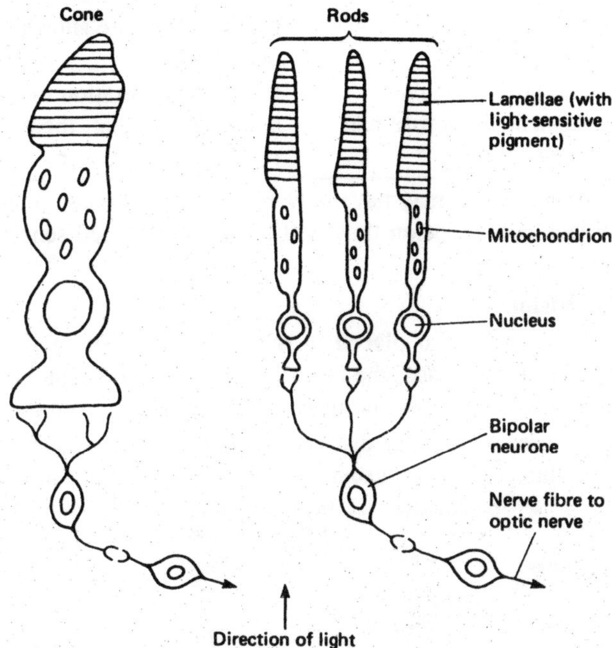

Figure 9.10: Cells of the retina.

Reflection of the light, once it has passed through the lens, is prevented by the dark, vascular choroid layer.

The light sensitive layer of the retina consists of two types of photoreceptors, rods and cones, containing pigments which are bleached by light. The bleaching process can initiate an action potential. There are certain differences between the two types of cell.

Table 9.8: Comparison of rods and cones

Rods	*Cones*
More numerous	Less numerous
Usually around periphery of retina	Usually located in centre of retina
Arranged in functional units served by one bipolar neurone, therefore acuity low	Each cone served by its own bipolar neurone, therefore acuity high
Very sensitive to low levels of illumination	Only stimulated by bright light
One type of rod only, stimulated by most wavelengths of visible light	Three types of cone, each selectively responsive to different wave

except red	lengths, therefore allowing colour perception
Rapid regeneration of light sensitive pigment, therefore can perceive flicker well	Slower regeneration of light sensitive pigment, therefore less responsive to flicker

Nocturnal animals may have only rods in their retinas, so although they may lack colour vision they will have a heightened sensitivity to movement.

Binocular Vision

The eyes may be so placed in the head that their two fields of vision overlap. In this case, the brain can use the two slightly different images it receives to assess the distance of the object from the retina.

Animals needing to judge distances accurately tend to have binocular vision like this, and it is usual in predators like cats and eagles.

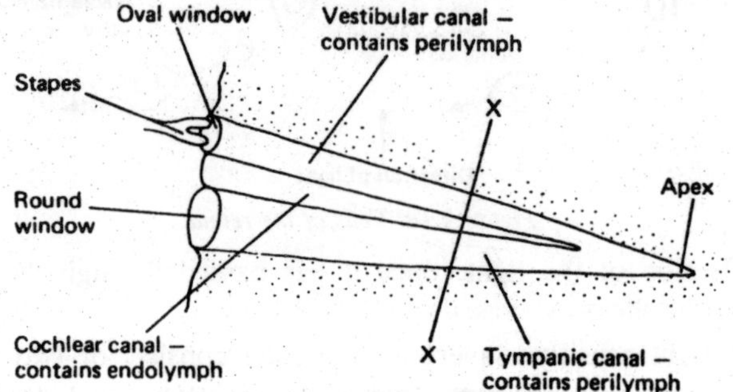

Figure 9.11: Uncoiled cochlea.

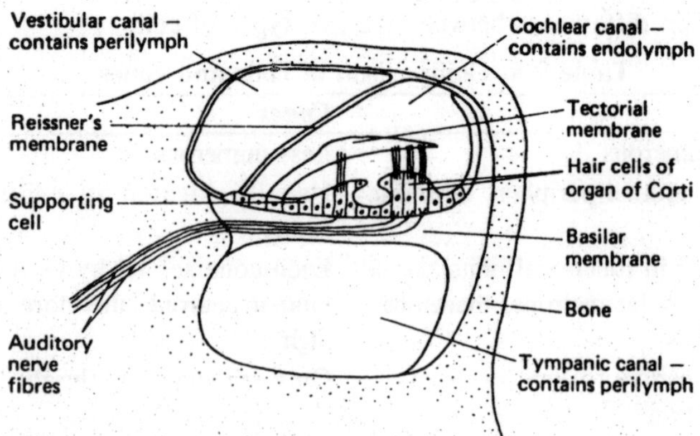

Figure 9.12: TS cochlea.

Humans have probably inherited their binocular vision from tree-climbing ancestors, who later developed the ability to throw spears.

Animals whose fields of vision do not overlap can see over a much wider area without turning their heads. Grazing animals such as sheep and cows, others likely to be preyed on like small birds, and slow-moving ones such as tortoises, tend to fall into this category.

Mechanoreceptors

The Ear

The receptor cells in the ear are equipped with sensitive hairs which are believed to initiate action potentials when they are moved. The pressure changes of sound waves move the hairs on one type of receptor, while movements or changes in position of the head deflect them on other types.

Hearing

The pinna, which in most mammals can be directed towards the source of sound, channels the sound waves towards the drum, which is sunk well into the head down a bony shaft.

The vibrations are relayed across the air-filled middle ear by the ossicles, which amplify them at the same time. Muscles attached to the ossicles prevent excessive vibration in response to a very loud noise. The eustachian tube opening into the back of the throat helps to equalize air pressure in the outer and middle ears.

The vibrations enter the fluid-filled cochlea via the oval window. The round window beneath it also vibrates, thereby neutralizing pressure changes in the cochlear fluid. The vibrating basilar membrane transfers

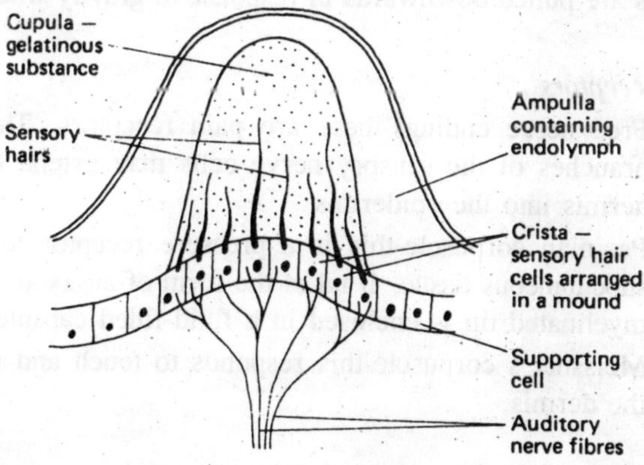

Figure 9.13: Ampulla.

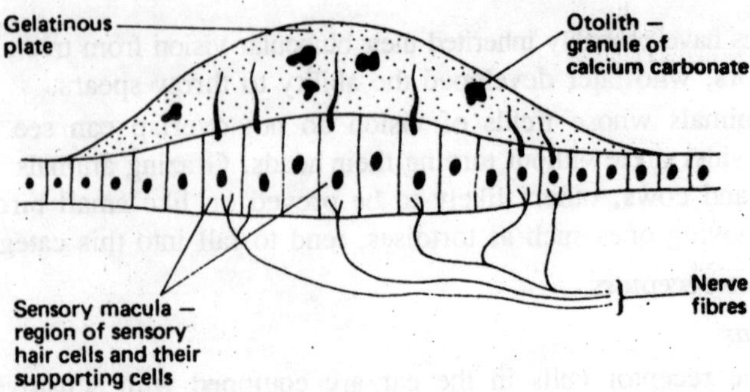

Figure 9.14: Utricle.

its movements to the sensitive hair cells of the organ of Corti, which press against the tectorial membrane above them, and an impulse is generated in the auditory nerve.

It is believed that cells in different regions of the organ of Corti respond to sounds of different pitch, those nearer the oval window being more sensitive to high notes than those further away from it.

Balance

Head movements are registered by the ampullae inside the semi-circular canals. The three canals are arranged at right angles to each other so that head movements in one direction will cause corresponding movements in the fluid inside the canals. The gelatinous cupula of the ampulla will be deflected by the moving fluid, and the hairs on the receptor cells will be moved.

Positions of the head are monitored by sensory hair cells in the utricule and the saccule. Otoliths of calcium carbonate balanced on these cells are pulled downwards in response to gravity when the head is tilted.

Dermal receptors

(i) Free nerve endings-these are pain receptors. The terminal branches of the sensory nerve cells may extend beyond the dermis into the epidermis

(ii) Pacinian corpuscle-this is a pressure receptor found in the subcutaneous tissue. It takes the form of an axon whose non-myelinated tip is enclosed in a fluid-filled capsule.

(iii) Meissner's corpuscle-this responds to touch and is found in the dermis.

10

LIFE CYCLE

TYPES OF REPRODUCTION AND LIFE CYCLES

Underlying Principles

No living organism can continue indefinitely even if it escapes predators, disease and other potentially fatal circumstances; therefore reproduction is essential for the survival of the species. Asexual reproduction, which is rapid and produces many offspring, is the best method for increasing the numbers of a species, but it suffers the disadvantage of producing identical offspring which have no scope for adapting to new situations.

Most life cycles therefore include meiosis, which produces some variety of offspring. In addition, many organisms involve two parents in a sexual process to further increase variety by the mixing of two genotypes.

Life cycles are adapted to the organism's mode of life; resistant phases are used to overcome adverse conditions. Larval stages are used in parasites to invade intermediate hosts as a means of returning to the primary host, and in free-living organisms as a means of exploiting food supplies and aiding dispersal.

Further adaptations of life cycles occurred in plants as they became terrestrial, with a shift from a dominant (water dependent) gametophyte to a dominant sporophyte able to withstand desiccation.

Points of Perspective

Knowledge of the functional significance of larvae in life cycles would be an advantage. Changes in life cycles associated with the evolutionary change from aquatic to terrestrial environment should be considered.

Essential Information

Asexual reproduction

This involves only one organism. The individuals produced are genetically identical, i.e. belong to a clone. Asexual-reproduction ensures rapid multiplication. There are five broad types of asexual reproduction: fission; budding; fragmentation; sporulation; vegetative propagation.

Fission

This is typical of bacteria and protists. The cell divides into two or more equal parts. In *binary fission* (splitting in two) growth of the population is exponential, i.e. one cell divides into two, two into four, four into eight, etc.

This is often a very rapid form of reproduction, e.g. many bacteria divide every 20 minutes. In many parasitic protista *multiple fission* occurs giving an even greater rate of reproduction, e.g. in the malarial parasite *Plasmodium,* where the nucleus divides into thousands of parts. This splitting is called schizogamy and each resultant cell is a schizont.

Budding

The parent produces an outgrowth, or bud, which detaches to become a separate individual, e.g. in *Saccharomyces* (yeast), *Hydra, Obelia.*

Fragmentation

This only occurs in simple organisms where the tissue is relatively undifferentiated, e.g. sponges and filamentous green algae, where, if small parts break off an organism, they will form whole new organisms.

Sporulation

This is the formation of small unicellular bodies which detach from the parent and, given suitable conditions, grow into new organisms. Spores are formed by bacteria, protozoa and many lower plants. Spore-bearing structures and spores vary but the spores are generally small, light, easily dispersed, have resistant walls and are produced in vast numbers.

Vegetative Propagation

Part of a plant becomes detached and grows into a new plant, e.g. the leaves of African violet develop into new plants. Some organs of vegetative propagation are also known as perennating organs since they enable the plants to survive adverse conditions, e.g. the development of a stem into a corm (crocus) or stem tuber (potato); the swelling of a root in carrot or dahlia; the swollen buds of bulbs such as onion.

Sexual Reproduction

This involves the fusion of specialized haploid cells called *gametes* which generally arise from two individuals. Usually the gametes differ in structure, size and behaviour (heterogametes), one being small, highly motile and produced in large numbers, the other being larger, non-motile and produced in small numbers.

Some lower organisms, e.g. certain algae and fungi, produce identical gametes (isogametes). Fusion of the gametes is called *syngamy*.

Table 10.1: Differences between asexual and sexual reproduction.

Asexual	*Sexual*
No mixing of genetic material; therefore less variation in offspring, therefore less evolutionary potential	Genetic mixing; therefore increased variation and great evolutionary potential
No gametes	Gametes
Usually more offspring	Fewer offspring
One parent	Usually two parents
Rapid; therefore takes advantage of favourable conditions	Longer process
May be resistant phases, e.g. spores, perennating organs	Rarely special resistant phase; may overcome difficult periods by delayed implantation (e.g. bats) or prolonged gestation (e.g. beavers)
The structures produced in asexual reproduction are more often used as a means of dispersal in animals than in plants	The structures produced are more often used as a means of dispersal in plants than in animals

Parthenogenesis

This is the development of a new organism from an unfertilized egg. Sometimes this egg has been produced by meiosis and is therefore haploid; it develops into the new individual e.g. drones in honeybee colonies.

Aphids produced by parthenogenesis are diploid because the eggs were produced by mitosis not meiosis.

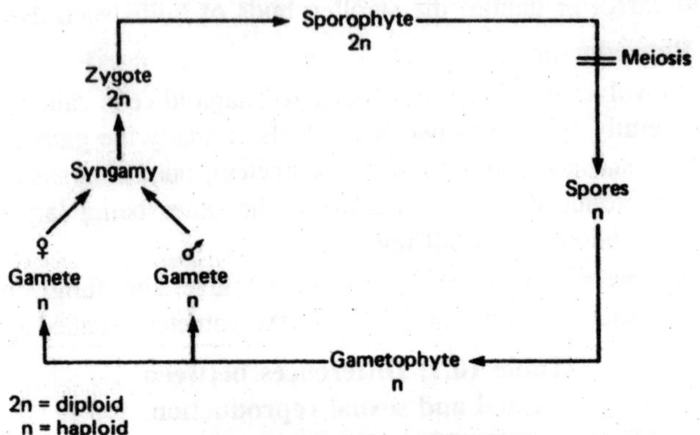

Many plants show a regular alternation of generations between a haploid gametophyte and a diploid sporophyte.

The relative significance of each stage in the life cycle varies with species.

(a) Generalized plant life cycle

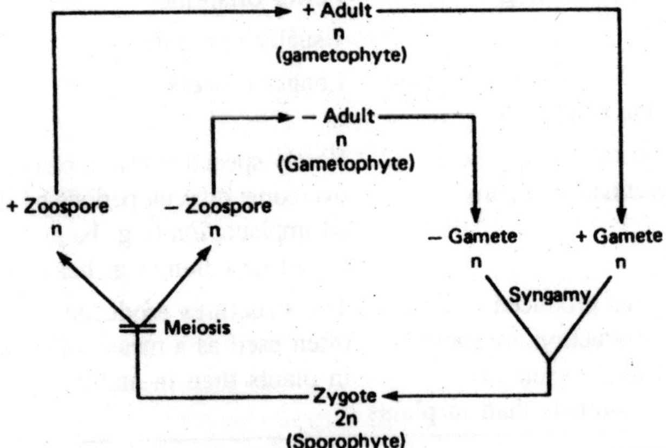

The zygote is the only diploid cell and meiosis occurs during its germination. The haploid phase multiplies extensively by mitosis giving a large number of unicellular individuals.

(b) Algae e.g. *Chlamydomonas*

Figure 10.1: Life cycles in plants. (Figure contd.)

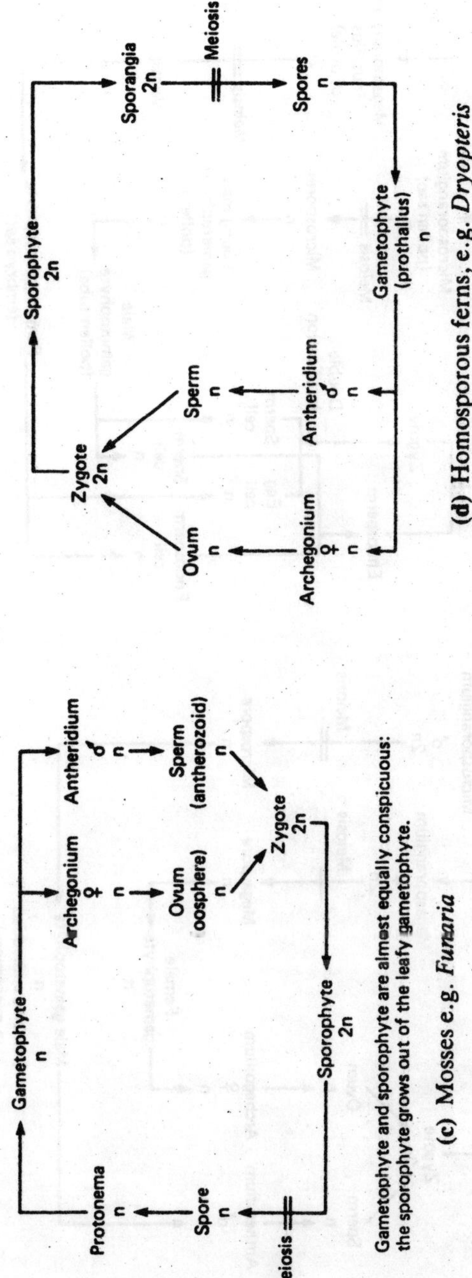

(c) Mosses e.g. *Funaria*

(d) Homosporous ferns, e.g. *Dryopteris*

Figure 10.1: Life cycles in plants. (Figure contd.)

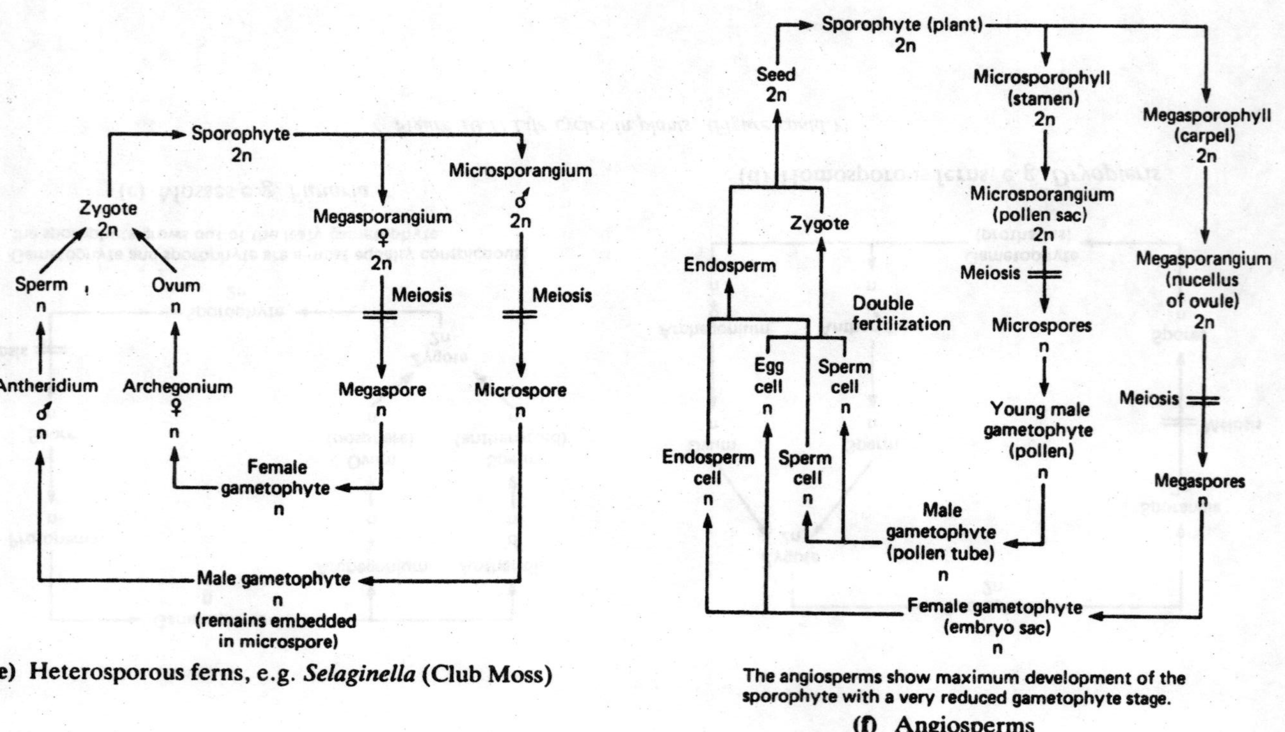

(e) Heterosporous ferns, e.g. *Selaginella* (Club Moss)

The angiosperms show maximum development of the sporophyte with a very reduced gametophyte stage.

(f) Angiosperms

Figure 10.1: Life cycles in plants. (Figure contd.)

All cells of multicellular animals are diploid except the gametes which are haploid.

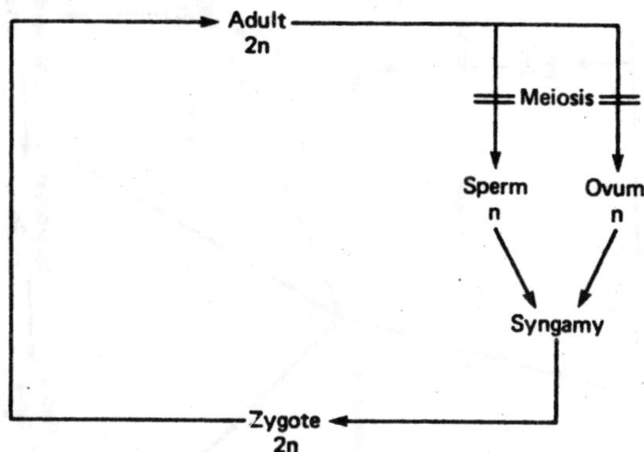

There is no true alternation of generations in animals.

(a) Generalized animal life cycle

Some animals do not show sexual reproduction, e.g. *Amoeba*.

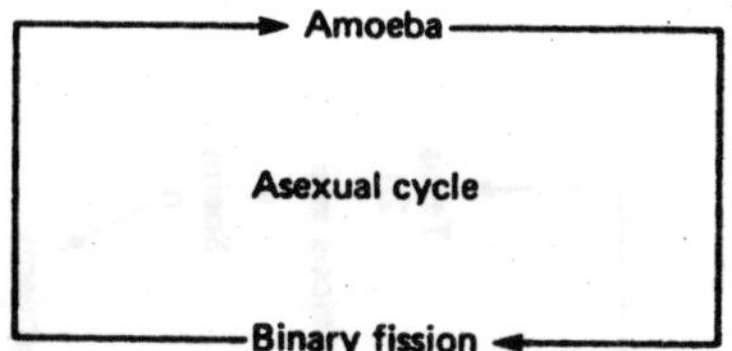

(b) Protozoans, e.g. *Amoeba*

Figure 10.2: Life cycles in animals. (Figure Contd.)

REPRODUCTION IN MAMMALS

Underlying Principles

Reproduction is concerned with the perpetuation of a species. Any mechanisms that increase the chance of fertilization and survival of the offspring will be of evolutionary advantage. Mammals owe much of their evolutionary success to the development of such mechanisms, which include:

1. Development of secondary sex characteristics to allow sexually mature individuals to recognize and mate with each other.
2. Seasonal breeding cycles that restrict copulation to times that will ensure birth at seasons most favourable to the survival of offspring.

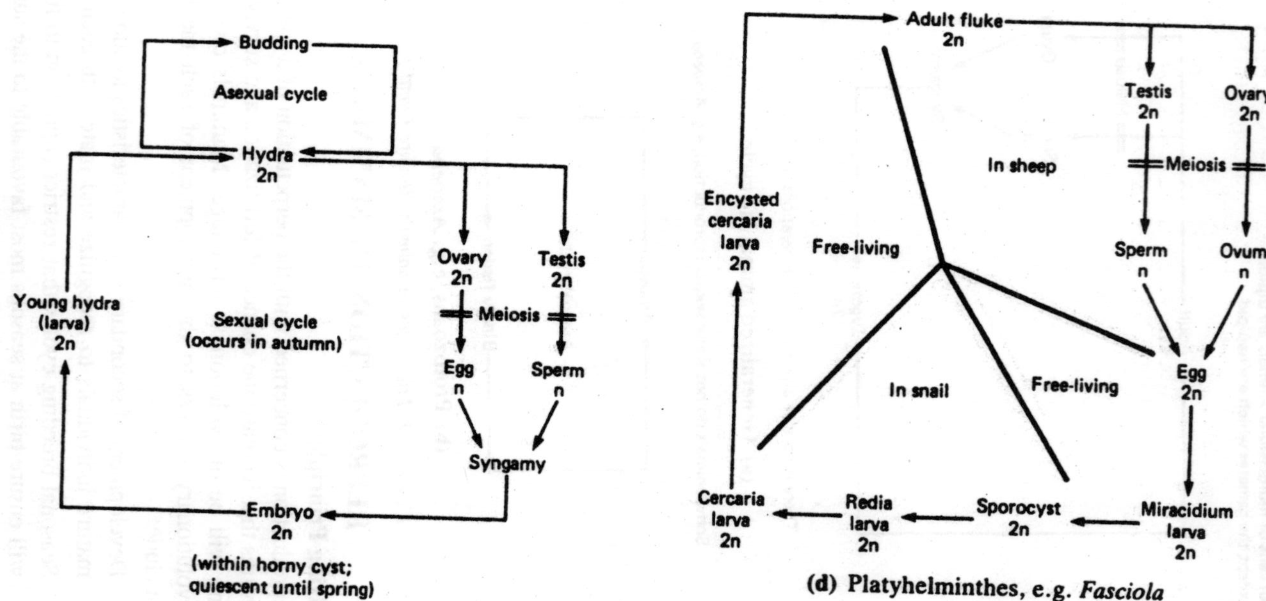

(c) Coelenterates, e.g. *Hydra*

(d) Platyhelminthes, e.g. *Fasciola*

Figure 10.2: Life cycles in animals. (Figure Contd.)

(a) Incomplete metamorphosis, e.g. dragonfly.

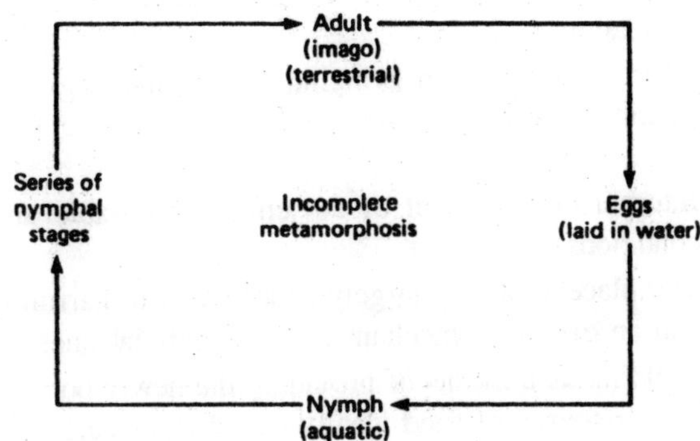

The nymph differs from the adult mainly in the absence of wings.

(e) Insects

(b) Complete metamorphosis, e.g. cabbage white butterfly.

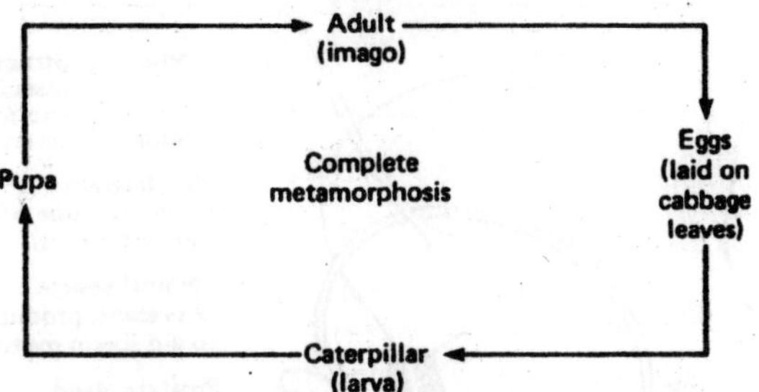

The larva is markedly different from the adult.

(e) Insects

Metamorphosis is also shown in the life cycles of some chordates, e.g. frog.

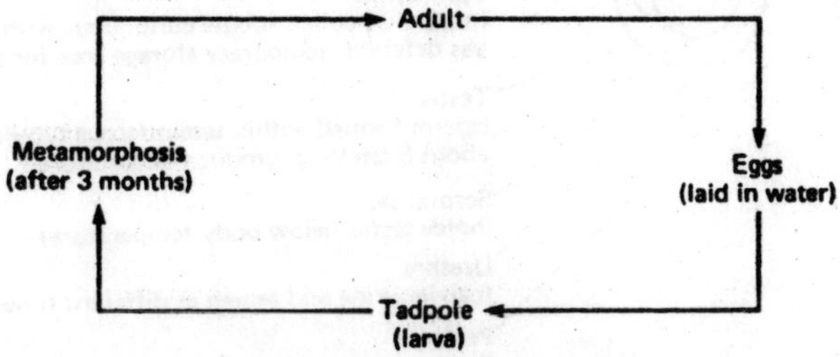

(f) Chordates, e.g. frog

Figure 10.2: Life cycles in animals. (Figure Contd.)

3. Female receptiveness to the male only when ovulation is taking place, or even ovulation being stimulated by the act of copulation.
4. Internal fertilization bringing sperm and egg close together within the relative safety and stability of the female genital tract.
5. Internal development of the embryo in stable and protected conditions.
6. The placenta acting largely as a barrier to harmful substances and an exchange mechanism for beneficial ones.
7. Suckling as a means of providing the newly born with a fairly secure source of food ideally suited to its early development.
8. Parental care allowing development of the young in controlled and protected conditions with the maximum use of learned behaviour, which has the advantage of being adaptable to meet varying circumstances.

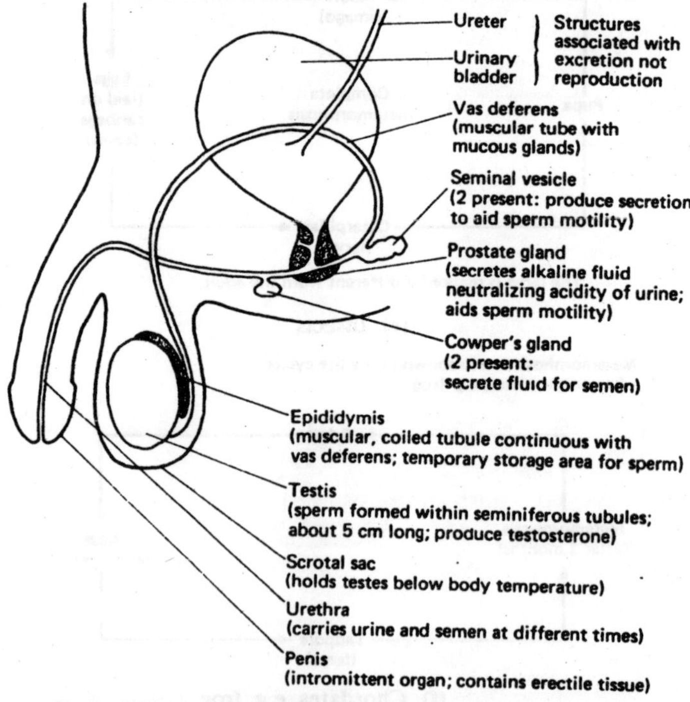

Figure 10.3 (a): Male reproductive system, lateral view (human).

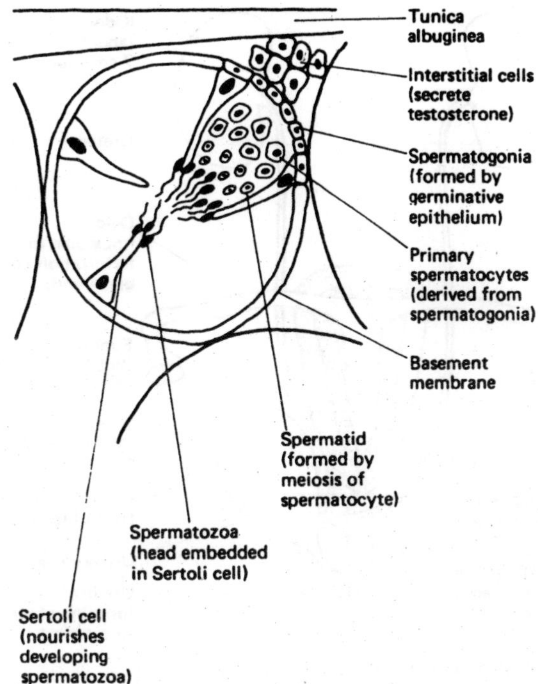

Figure 10.3 (b): TS seminiferous tubule (Each testis contains about 1000).

Points of Perspective

A knowledge of the biological basis for the various methods of birth control would be an advantage.

It is essential to know about the reproductive processes of mammals other than man, if only to illustrate those points not applicable to humans, e.g. delayed implantation, yearly oestrous cycles and copulation inducing ovulation.

The importance of an understanding of breeding cycles in mammals to the development of animal husbandry should be appreciated, in particular in respect of man's domesticated animals.

A background knowledge of the medical aspects of reproduction would be useful, including the biological significance of ante- and post-natal care.

Essential Information

Sexual reproduction involves the fusion of two haploid gametes to form a zygote. The male gamete (spermatozoon) is small, highly motile and produced in large numbers whereas the female gamete (egg cell)

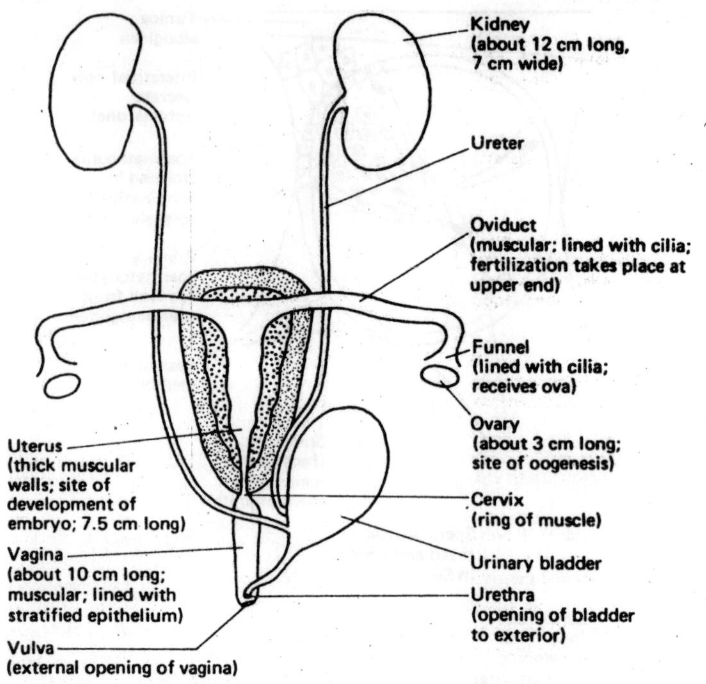

Figue 10.3 (c): Female urinogenital system, ventral view (human).

is larger, non-motile and fewer are produced. Egg cells, or ova, are produced in the ovary by a process known as oogenesis; spermatozoa are produced in the testes by spermatogenesis.

Fertilization

When the head of the spermatozoon comes into contact with the vitelline membrane, the acrosome opens releasing a chemical which softens the membrane and the inner membrane of the acrosome inverts to form a filament which pierces the egg membranes.

This process is called the acrosome reaction. Following this the cortical granules apply themselves to the inner surface of the vitelline membrane, thickening it and preventing the entry of further sperm.

The head and middle piece move through the cytoplasm and the nuclei of the sperm and egg cell fuse, thus restoring the diploid number.

Development of the zygote

The zygote moves down the oviduct to the uterus, a journey which takes about one week in humans. By the time it has reached the uterus it has divided mitotically to form a hollow ball of cells—a

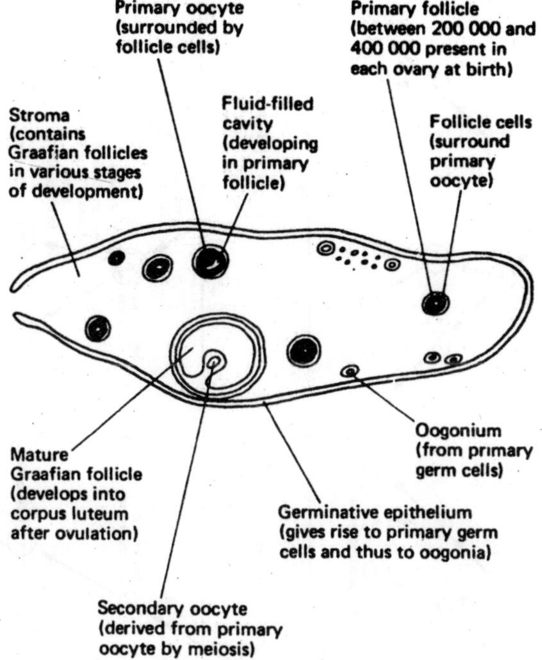

Figure 10.3 (d): LS ovary.

blastocyst. The blastocyst becomes implanted in the lining of the uterus and the outer layer of cells forms trophoblastic villi which project into the uterine wall.

As the embryo develops it becomes surrounded by a series of extra-embryonic membranes: the amnion, which encloses a fluid-filled cavity which eventually fills the entire uterus, and the allantois and chorion, which unite to form the allanto-chorion which develops into the placenta.

The only connection between the embryo and the wall of the uterus is the stalk of the allantois which becomes the umbilical cord. This contains the umbilical artery and vein conveying foetal blood to and from the placenta.

The placenta

Exchange of materials takes place between the maternal blood spaces in the uterine wall and the foetal capillaries in the chorionic villi.

The functions of the placenta

1. Oxygen, water, soluble food and salts pass from maternal to

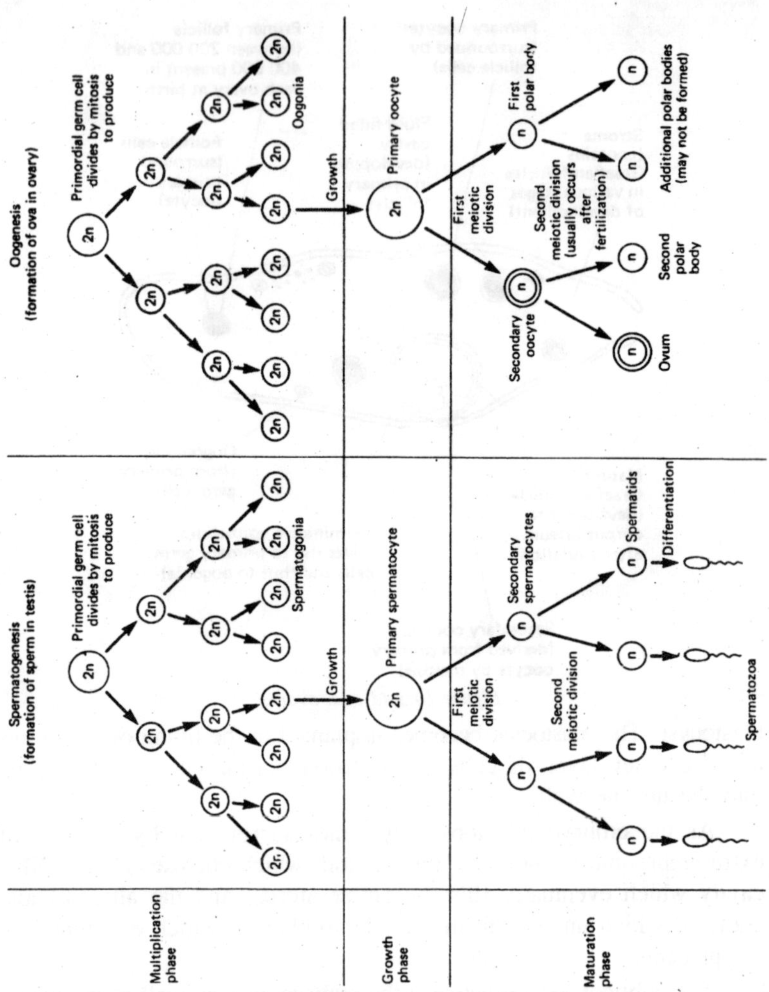

Figure 10.4 (a): Gametogenesis

foetal blood.

2. Carbon dioxide and nitrogenous waste pass from foetal blood to maternal blood for removal by the mother.
3. During processes 1 and 2 it prevents mixing of foetal and maternal bloods since the foetus may inherit the father's group which, if incompatible with the mother's, would damage the foetus if mixing occurred.
4. During pregnancy the placenta progressively takes over the role of producing the hormones that prevent ovulation and

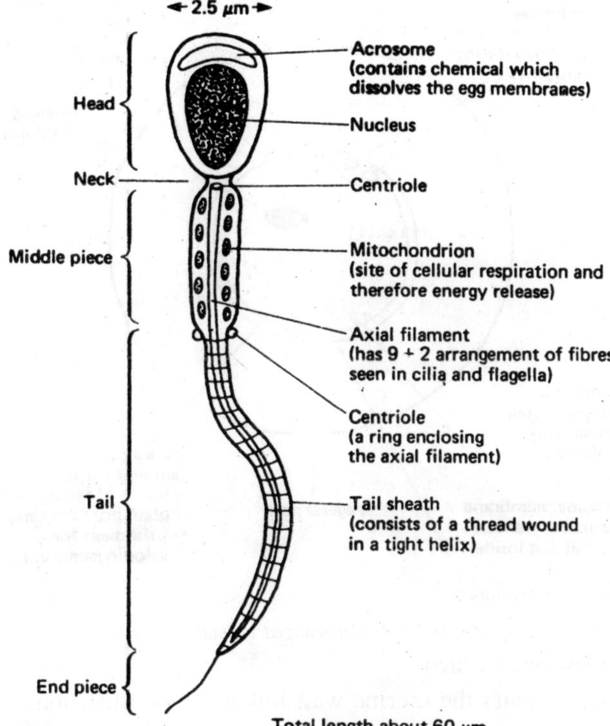

Figure 10.4 (b): Human spermatozoon, based on electron micrograph.

menstruation.

5. It allows maternal and foetal blood pressures to differ from one another.
6. It prevents the passage of some pathogens from the mother to the foetus, although some viruses and bacteria and many antibodies can cross between the two.
7. While allowing some hormones across, it prevents the passage of those maternal hormones that could adversely affect foetal development.

Female sexual cycle (oestrous or menstrual cycle) In humans this is a twenty-eight day cycle controlled by hormones secreted by the pituitary gland and the ovary. The main hormones involved are:

(a) Follicle stimulating hormone (FSH) which
  (i) causes Graafian follicles to develop in the ovary.
  (ii) stimulates the tissues of the ovary to produce oestrogen.

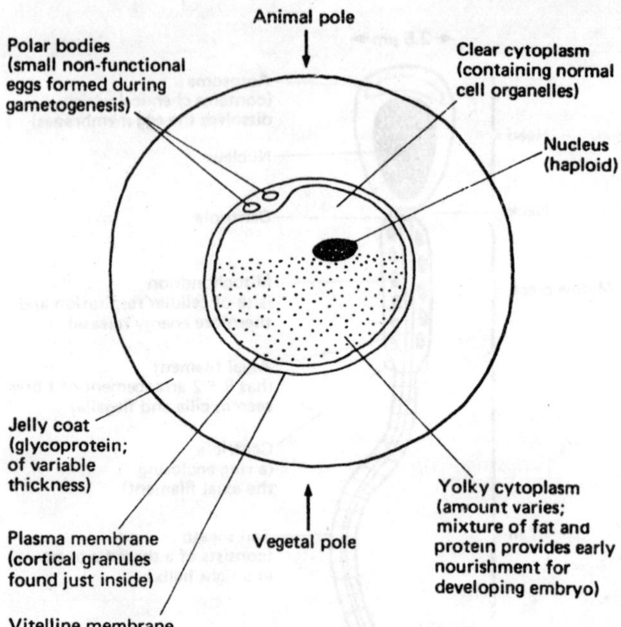

Figure 10.4 (c): Generalized animal egg cell.

(b) Oestrogen which
   (i) repairs the uterine wall following menstruation.
   (ii) builds up in concentration during the first two weeks of the menstrual cycle until it stimulates the pituitary gland to produce luteinizing hormone.

(c) Luteinizing hormone (LH) which
   (i) brings about ovulation.
   (ii) causes a Graafian follicle to develop into a corpus luteum which produces progesterone.

(d) Progesterone which
   (i) inhibits FSH production and therefore stops further follicles developing (a fact made use of in the development of the contraceptive pill).
   (ii) causes development of uterine wall prior to implantation.

Male Sexual Cycle

There is no cycle in male humans but the gonads are regulated by hormones identical with those of the female. FSH, which is generally called interstitial cell stimulating hormone (ICSH) in the male, promotes

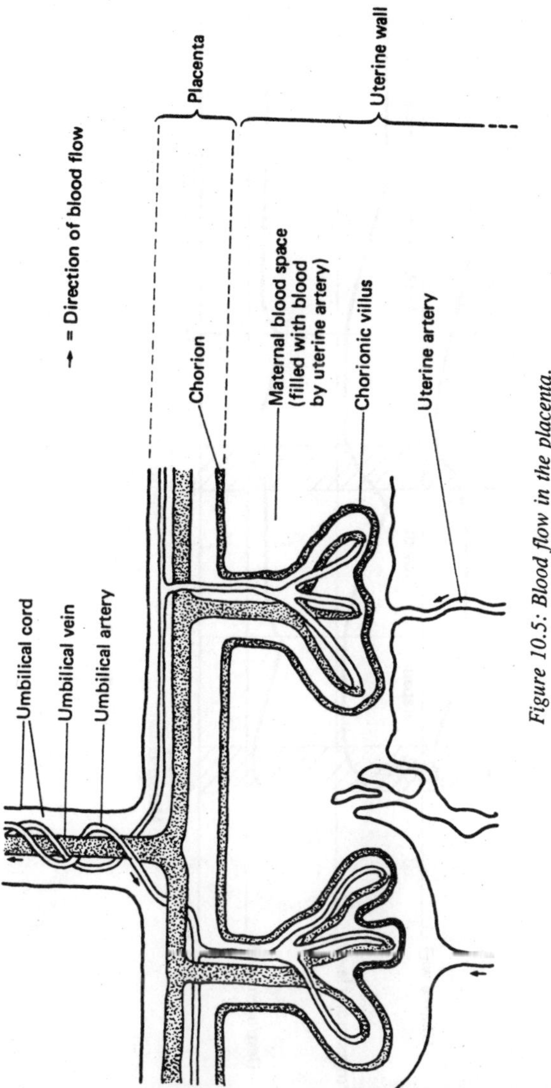

Figure 10.5: Blood flow in the placenta.

spermatogenesis. LH, also called ICSH, causes the interstitial cells between the seminiferous tubules to secrete androgens, hormones which stimulate the development of male secondary sex characteristics.

Breeding cycles in vertebrates

If the cycle is seasonal, environmental changes trigger the endocrine system at the most advantageous time of the year for breeding.

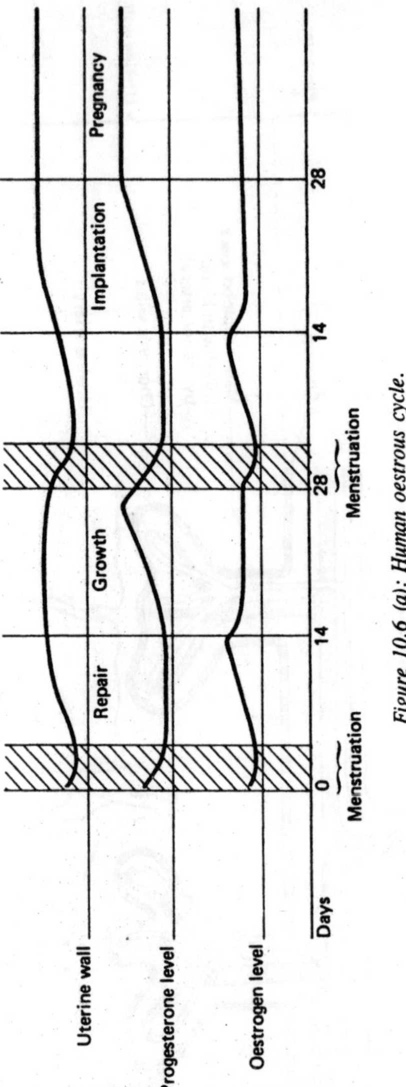

Figure 10.6 (a): Human oestrous cycle.

Whereas in the stickleback *Gasterosteus aculeatus* daylength is the most important environmental trigger and temperature plays a subsidiary role, in the minnow *Couesius plumbeus* the cycle is most affected by temperature.

The breeding of birds is nearly always seasonal. In temperate latitudes increase in daylength in the spring acts through the pituitary gland to cause ripening of the gonads. Changes in behaviour associated

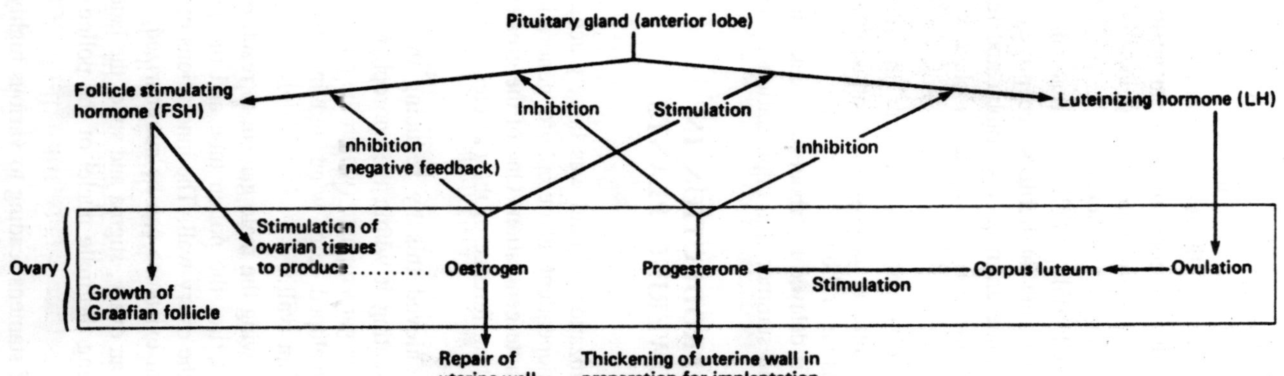

Figure 10.6 (b): Hormonal control of the oestrous cycle.

with the development of the gonads vary with species. Other factors such as degree of activity and food eaten may be important, especially near the equator where there is little seasonal variation.

If the cycle is more or less continuous, as in most endotherms and some tropical ectotherms, the controls are usually within the organism and rely on a feedback mechanism.

Usually the female will only receive the male during a strict period of oestrus (or heat) but in humans receptivity may occur throughout the cycle. In some animals, e.g. rabbit and ferret, coitus induces ovulation and fertilization usually takes place within a few hours.

In certain bats copulation occurs in autumn and fertilization is delayed until spring. Implantation usually occurs about one week after fertilization, as in humans and rabbits, but in some animals, e.g. pig, cat and dog, implantation is delayed for about two weeks and in badgers mating takes place in the summer but implantation is delayed until late in the winter.

REPRODUCTION IN FLOWERING PLANTS

Underlying Principles

The successful colonization of land was largely achieved by the reduction of the waterdependent gametophyte generation and the dominance of the sporophyte generation. One of the greatest problems to be overcome was the transference of the gametes in the absence of water.

The angiosperms achieved this by reducing the gametophyte generation and then protecting it within the sporophyte. Methods of transferring the young male gametophyte (within the pollen grain) from one plant to another developed, e.g. wind and insects, while it is protected within a resistant wall.

A mechanism for allowing the male gamete to reach the protected female gamete developed, i.e. the pollen tube and the protection of the resulting embryo by the ovary wall. The innovations of angiosperm reproduction to terrestrial existence can be summarized:

1. Development of an ovary, stigma and style, the latter providing a pathway guiding the male nuclei of the pollen tube to the embryo sac.
2. Development of stamens leading to various highly successful systems of pollination.

3. Reduction of the female gametophyte to a 7-celled embryo sac and enclosure within the ovary for protection.
4. Reduction of the male gametophyte to a pollen tube with the complete elimination of motile sperm.
5. Presence of a double fertilization resulting in the embryo and the endosperm. The latter is contained within the seed and forms the source of nourishment for the young embryo.
6. Development of the fruit, usually from the ovary wall.

Points of Perspective

It would be useful to have knowledge of a wide range of pollination mechanisms with precise details of the adaptations of one flower example to each type.

The importance of the parallel evolution of insects and flowering plants should be appreciated.

The storage materials in seeds and the experimental tests for them as well as their economic importance as human food sources, e.g. cereals, peas, bean, soya, should be known.

A good candidate should know some details of the length of dormancy in seeds and the factors that induce and break it.

Some outline knowledge of the different types of fruit, e.g. dry, succulent and false fruits, would be helpful.

The candidate with a good detailed knowledge of a number of methods of dispersal, especially some of the less usual ones, e.g. squirting cucumber, will obviously have an advantage.

The candidate should have performed, or know how to perform, laboratory experiments to determine (i) factors affecting germination, (ii) the respiratory rate (RQ) of germinating seeds and (iii) the enzymatic activity of seeds, e.g. amylase action.

Essential Information

A flower is a vegetative shoot whose parts are adapted for reproduction. The female part of the flower is referred to as the gynoecium and consists of the carpels. The male part is the androecium and consists of the stamens. The other parts of the flower are mainly concerned with protection and the attraction of insects.

Production of Pollen

The centre of the pollen sacs contains pollen mother cells each of which divides meiotically to give a tetrad (four) of pollen grains. These four cells later separate. Each young grain has one nucleus but

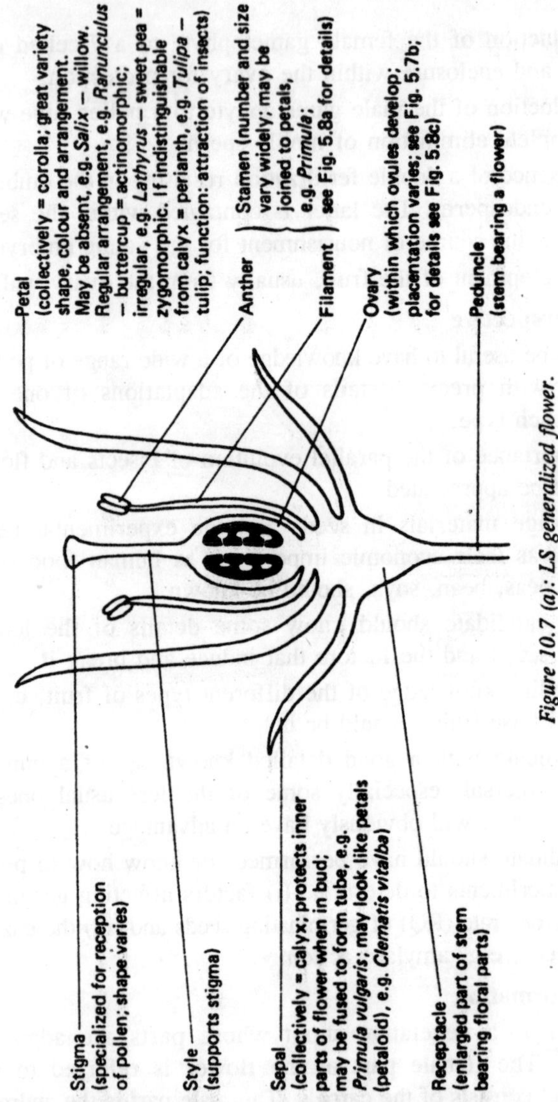

Figure 10.7 (a): LS generalized flower.

this divides into two. One of these, surrounded by denser cytoplasm, forms the generative cell and this later gives rise to the male gametes; the other forms the tube nucleus.

The pollen grain has a double wall, a delicate inner one of cellulose (the intine) and an outer one of variable thickness (the exine). Pollen grains vary in shape and in size from 3-300 μm. They are

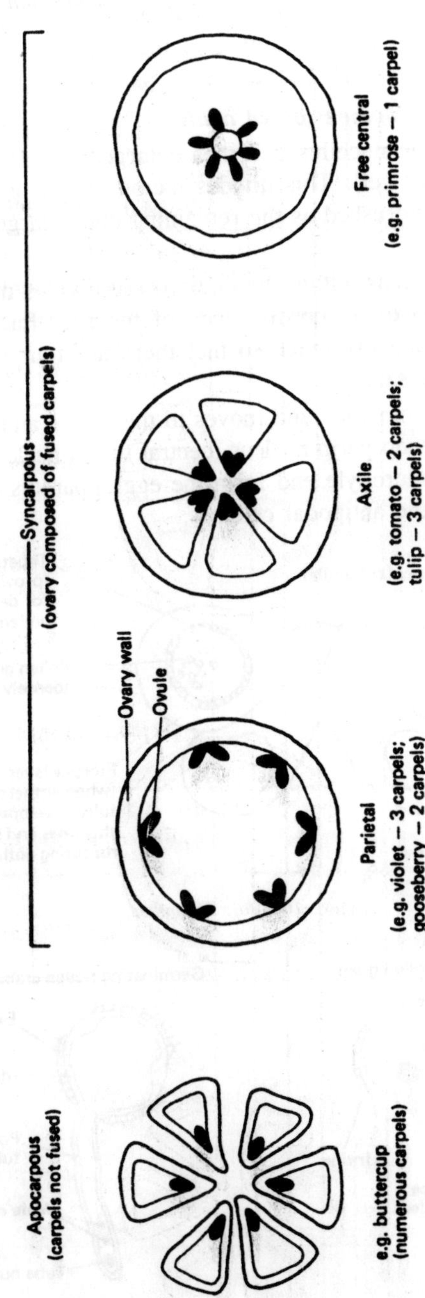

Figure 10.7 (b): Common forms of placentation.

liberated by the longitudinal rupture of the anther lobe together with the breakdown of the wall between each pair of sacs.

Structure and development of an ovule

One cell of the nucellus becomes enlarged and is known as the embryo sac mother cell. This divides meiotically to give four cells, three of which are crushed as the remaining one enlarges to form the embryo sac.

The single nucleus within the embryo sac divides mitotically and the two nuclei move to opposite ends of the sac. Each of the two nuclei divides mitotically twice so that there are four haploid nuclei at each end of the sac.

One nucleus from each end moves to the centre and these fuse to form the primary endosperm nucleus (central fusion nucleus). The three remaining at the micropyle end form the egg apparatus and the three at the other end, the antipodal cells.

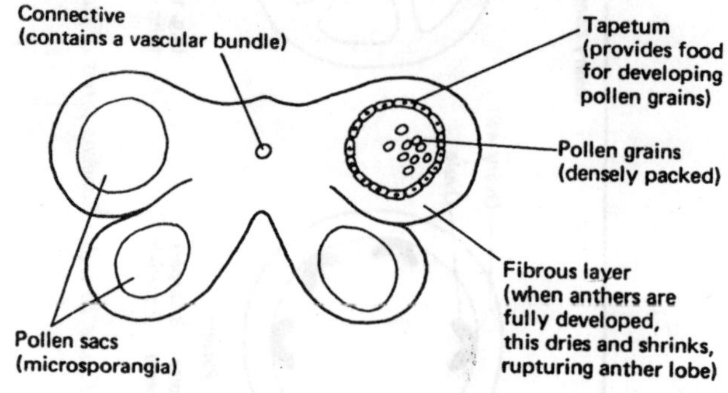

Figure 10.8 (a): TS anther.

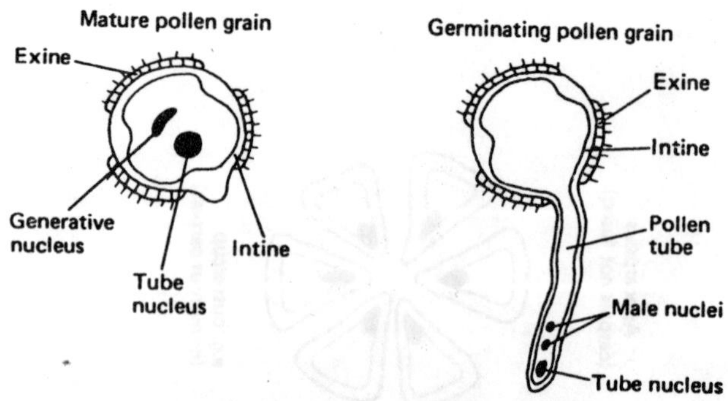

Figure 10.8 (b): Germination of pollen grains.

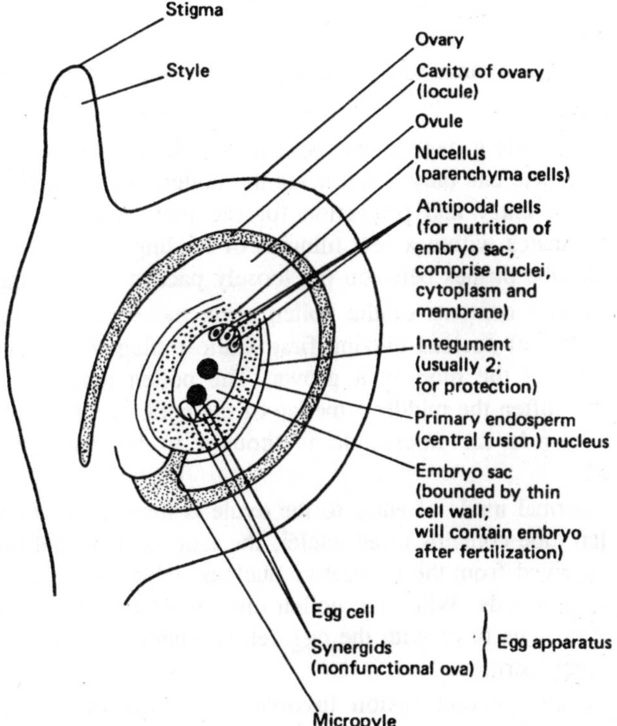

Figure 10.8 (c): LS generalized carpel.

Pollination

This is the transference of pollen from the anther to the stigma.

Self pollination

The transference of pollen from the anther to the stigma of the same flower, or a different flower on the same plant. Many plants are designed to self pollinate should cross pollination fail, e.g. in the Compositae as the flower ages the stigmas curl to touch the anthers.

A few species of plants produce flowers which become self pollinated while still in the bud stage (= cleistogamy) but this usually only occurs in a small proportion of the flowers, e.g. *Viola* (violet), where open flowers may also be self pollinated by insects.

Cross pollination

The transference of pollen from the anther to the stigma of a flower on a different plant of the same species. Although some plants regularly show self pollination there is a general tendency towards cross pollination which reduces inbreeding and increases the variability of a population.

The two main agents of cross pollination are wind and animals (mainly insects). The main characteristics of wind- and insect-pollinated flowers can be summarized as in Table elsewhere in this chapter.

Fertilization

As the female gamete in angiosperms is protected within the carpel, the male gamete can only reach it via the pollen tube, which provides a channel of entry and protection for the male nuclei. The pollen grain germinates within a few minutes of landing on the stigma and the pollen tube pushes between the loosely packed cells of the style.

The entire contents of the pollen grain move into the tube, the tube (vegetative) nucleus moving first. Callose plugs block the older, empty parts of the tube as it grows. The pollen tube may secrete pectases to soften the middle lamellae of the cells of the style and the growth towards the micropyle is thought to be chemotropically controlled.

The normal mode of entry to the ovule is through the micropyle. The pollen tube contains three nuclei, the tube nucleus and two male gametes derived from the generative nucleus, which keep near the tip as growth proceeds. When the pollen tube penetrates the embryo sac one male nucleus fuses with the egg cell (oosphere) and one with the primary endosperm nucleus.

Since this second fusion involves three nuclei (the primary endosperm nucleus was derived from two polar nuclei), it is called triple fusion. Double fertilization is said to occur because two male nuclei fuse with the female nuclei. The fertilized oosphere gives rise

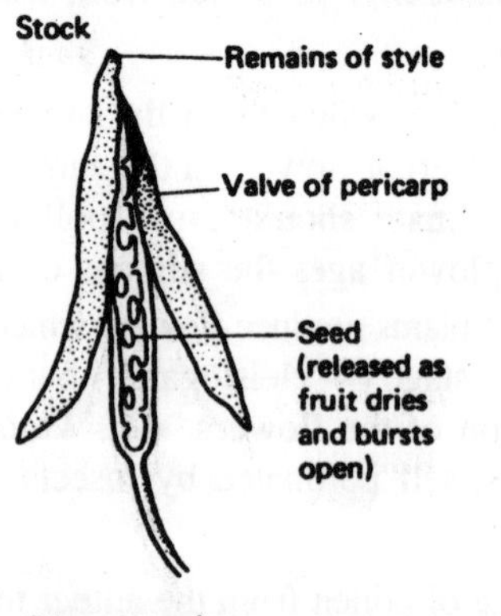

Figure 10.9 (a): Mechanically-dispersed fruits and seeds.

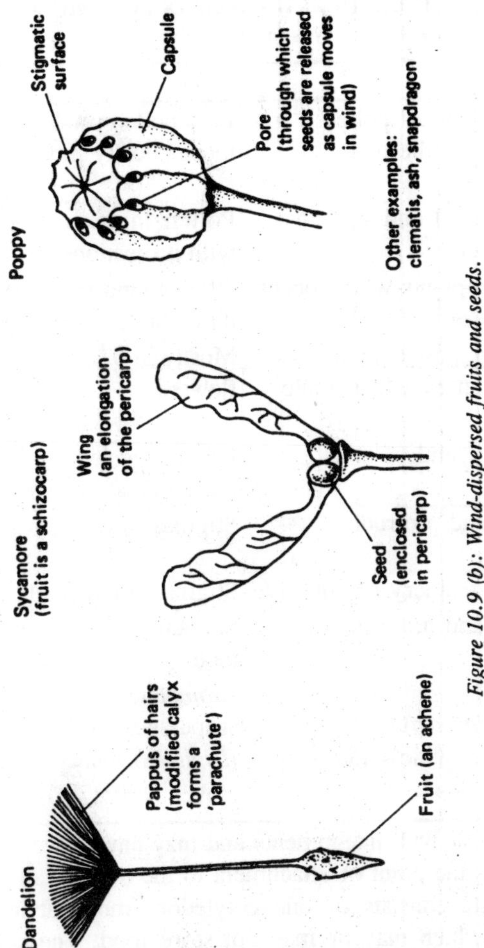

Figure 10.9 (b): Wind-dispersed fruits and seeds.

to the embryo and the fertilized primary endosperm nucleus to the endosperm.

In non-endospermic seeds the endosperm is used up to form the cotyledons before the seed is ripe but in endospermic seeds nuclear and cellular divisions of the fertilized primary endosperm nucleus give rise to an extensive endosperm.

Seeds and Fruits

The development of the fertilized ovule produces a seed and the ovary as a whole develops into the fruit.

The mature seed is covered by a hard leathery testa, which is the

Table 10.2: Differences between wind- and insect-pollinated flowers.

Wind	*Insects*
Enormous amounts of pollen produced because there is great wastage	Less pollen produced because mechanically more precise
Pollen small, light and smooth	Pollen larger and heavier often with projections
Often found in plants which occur in groups	Often found in plants which are more or less solitary
Often found in unisexual flowers (usually with an excess of male flowers)	Mostly in bisexual (hermaphrodite) flowers
Flowers dull, scentless and nectarless	Bright, scented flowers with nectar therefore attractive to insects
Stigmas long and protrude above petals	Stigmas often deep in corolla
Stigmas often feathery or adhesive	Stigmas often small
Stamens long and protrude above petals	Stamens may be within corolla tube
Examples: grasses (e.g. *Festuca);* hazel *(Corylus);* willow *(Salix);* plantain *(Plantago)*	*Examples:* snapdragon *(Antirrhinum);* foxglove *(Digitalis);* buttercup *(Ranunculus);* clover *(Trifolium)*

product of one or both integuments and may have a scar, or hilum, on it which marks the point of attachment to the ovary wall. The embryo within the testa consists of one cotyledon (monocotyledon) or two (dicotyledon) which may or may not store food. The cotyledons are attached to a central axis differentiated into a plumule (shoot) at one end and a radicle (root) at the other.

Examples of types of seeds:

dicotyledon, non-endospermic: broad bean, pea, sunflower dicotyledon, endospermic: castor oil monocotyledon, non-endospermic: water plantain monocotyledon, endospermic: maize, onion

Dispersal of Seeds and Fruits

Sexual reproduction ensures variability and a potential for the colonization of new habitats. If this potential is to be realized dispersal is essential. Dispersal also reduces the chances of backcrossing to the

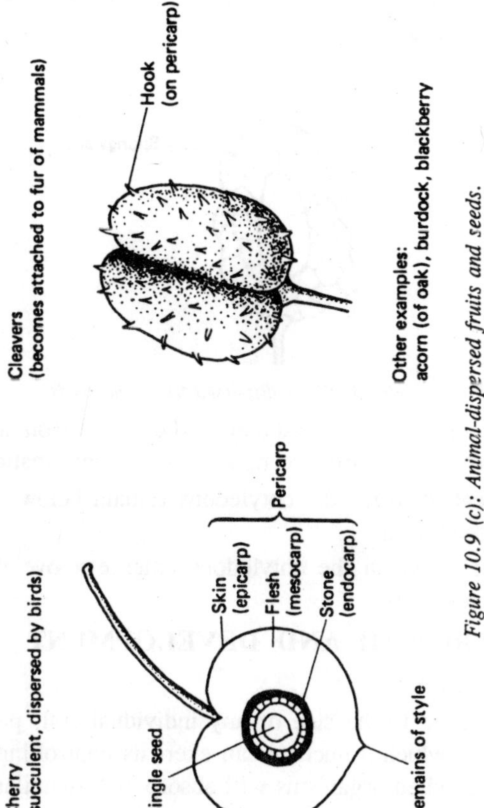

Figure 10.9 (c): Animal-dispersed fruits and seeds.

parents and the resultant problems of inbreeding. It also prevents overcrowding and competition and decreases vulnerability to epidemic attacks.

There are four main mechanisms of dispersal: mechanical, wind, animal and water.

Germination

A seed begins to develop after a period of dormancy and is dependent on the food reserves in the cotyledon until the first true leaves develop. Germination requires an adequate supply of oxygen and water and a suitable temperature.

As the first stage of germination is the absorption of water and since testas are frequently impermeable to water, germination may be delayed until the testa is broken or decayed. The water renews the physiological activities of the endosperm and embryonic tissues. The

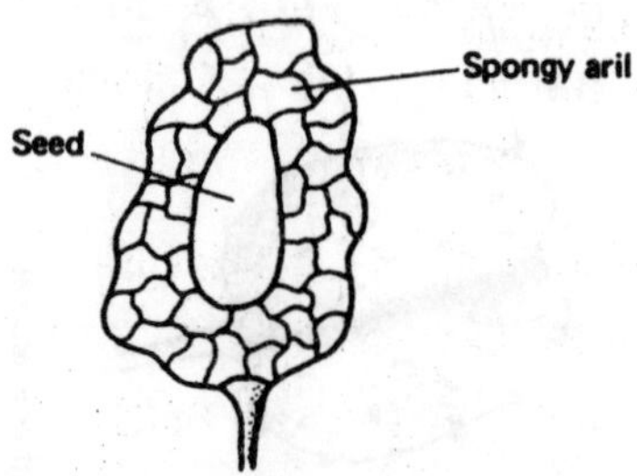

Figure 10.9 (d): Water-dispersed fruits and seeds.

enzymes in the seeds are activated and food changed from an insoluble to a soluble form. There are two main types of germination:

1. hypogeal in which the cotyledons remain below the ground, e.g. broad bean;
2. epigeal in which the cotyledons emerge above the ground, e.g. runner bean.

GROWTH AND DEVELOPMENT

Underlying Principles

There is a limit to the size of any individual cell, partly due to the distance over which a nucleus can exert its controlling influence. Therefore single celled organisms will absorb and assimilate materials until they reach a certain size, when they divide to give two separate individuals.

The multicellular condition allows organisms to attain greater size. This brings its own problems but these are largely outweighed by such advantages as:

1. increased scope for differentiation, specialization of function and hence greater efficiency.
2. the ability to separate more easily processes requiring different conditions, e.g. different pH values.
3. increased ability to store materials.
4. the ability to replace damaged cells from those remaining.
5. greater competitive advantage, e.g. large plants have better access to light than small ones.
6. some security from predators that find a large organism more difficult to capture and ingest.

All multicellular organisms originate as a single cell, the zygote, and undergo three phases of development:

1. Growth: an irreversible increase in mass.
2. Differentiation: the development of differing cell structures and functions.
3. Morphogenesis: the development of the overall form of organs and hence the organism.

In addition to the importance of growth and development in the early stages of an organism's life, these events continue to manifest themselves later in such processes as regeneration, repair and gametogenesis.

Points of Perspective

Knowledge of a wide range of methods for measuring growth would be helpful. It would be a considerable advantage to have carried out basic growth measurement experiments on a yeast colony, a stem or a root and a small mammal. In addition it would be useful to have carried out experiments to determine the effects of external factors such as temperature on the rate of growth.

It is necessary on some syllabuses to study the embryology of chosen examples by microscopic examination. Even where the syllabus does not require it, the candidate's understanding of the process would be improved by such a study.

Essential Information

Growth

Growth is an irreversible increase in size during development. It usually occurs in three distinct phases: cell division, cell assimilation and cell expansion.

Measurement of Growth

Using a single parameter for the measurement of growth may not take into account growth in all directions, e.g. measuring an increase in length does not take into account changes in girth. Changes in volume are difficult to measure in irregular organisms and changes in fresh weight may be complicated by temporary changes, e.g. drinking water.

Dry weight gives a less misleading picture but, since the measurement of this involves killing the organism, numerous similar organisms are required. A further complication is that an overall measurement of growth does not take into account the fact that different

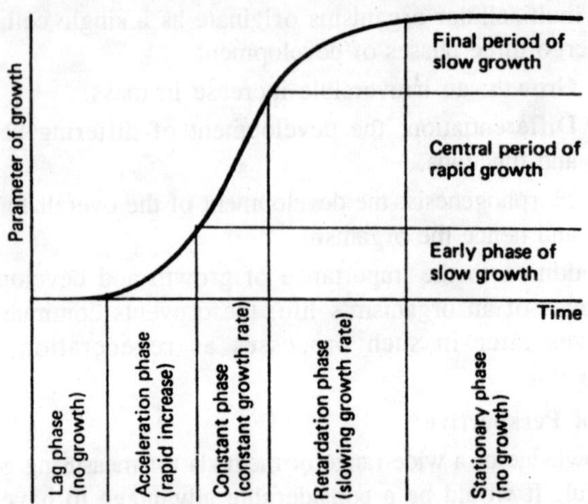

This sigmoid (S-shaped) curve can be applied to a population, an individual or an organ of an individual, although the pattern may be modified.

Figure 10.10 (a): General growth curve.

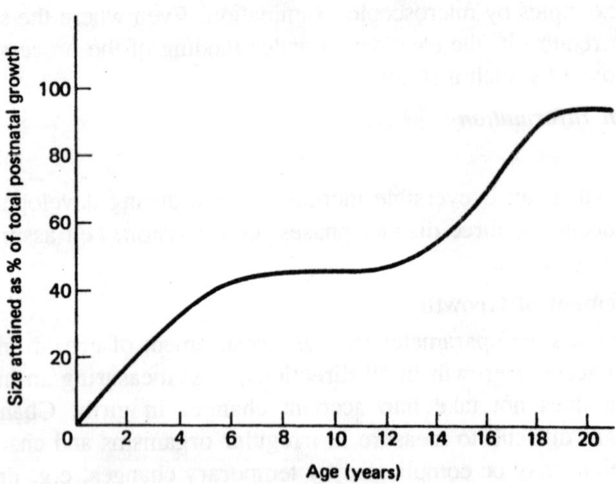

This resembles two flattened sigmoid curves on top of each other. The first represents the early growth phase in a child, the second the later growth phase in adolescence.

Figure 10.10 (b): Human growth curve.

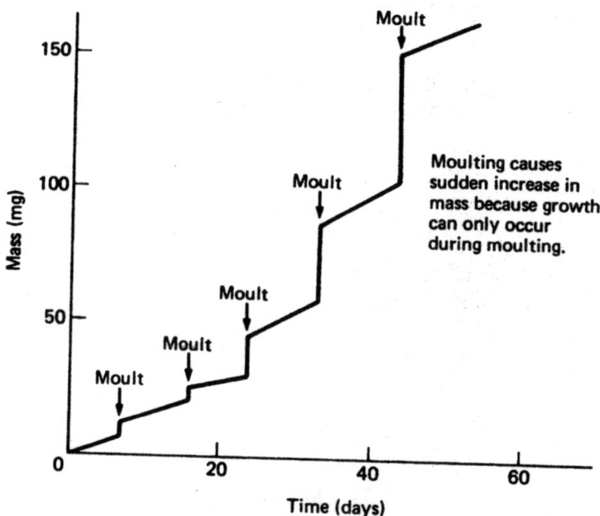

Figure 10.10 (c): Growth curve of an arthropod, e.g. Notonecta glauca.

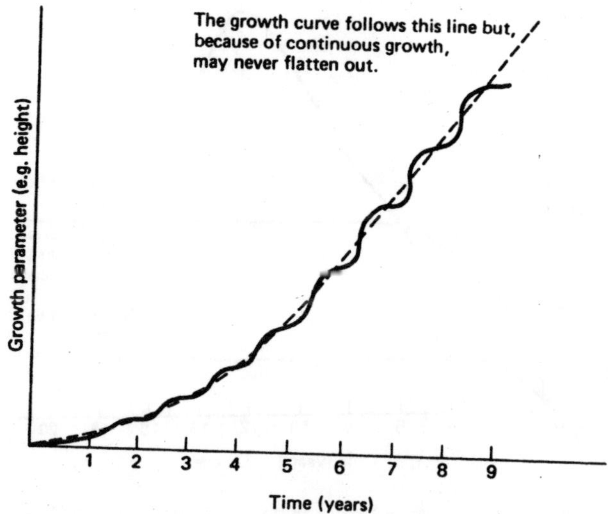

Figure 10.10 (d): General growth curve.

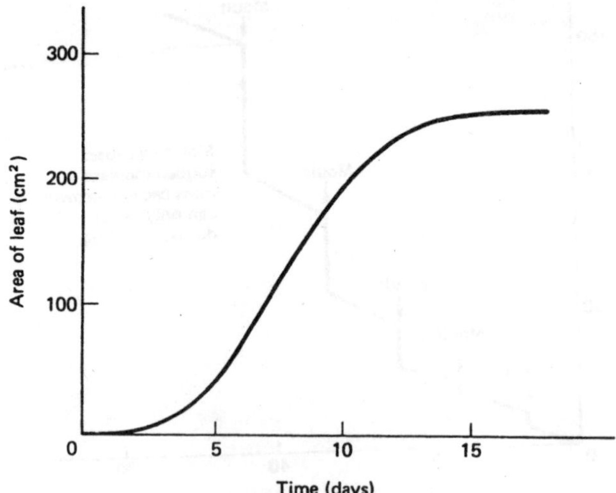

In plants growth of an organ often reflects the growth of the whole organism.

Figure 10.10 (e): Growth in area of a cucumber leaf

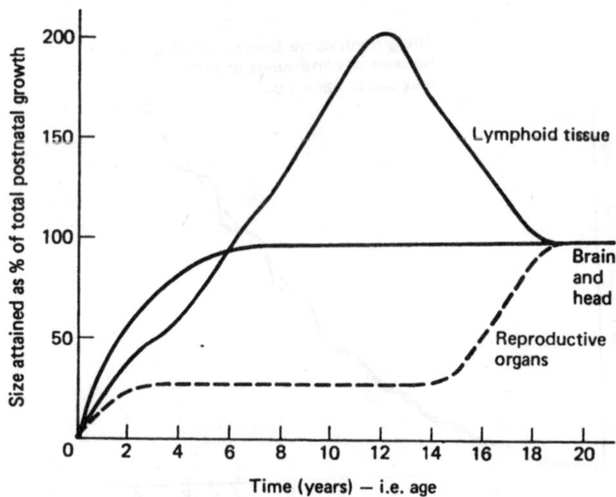

In animals growth of organs is often allometric.

Figure 10.10 (f): Growth curves of some human organs

organs may have their own peculiar growth rates (allometric growth).

Growth curves

Figure elsewhere in this chapter represents a variety of growth curves in plants and animals.

Growth rate

A graph of growth rate is drawn by plotting growth increments

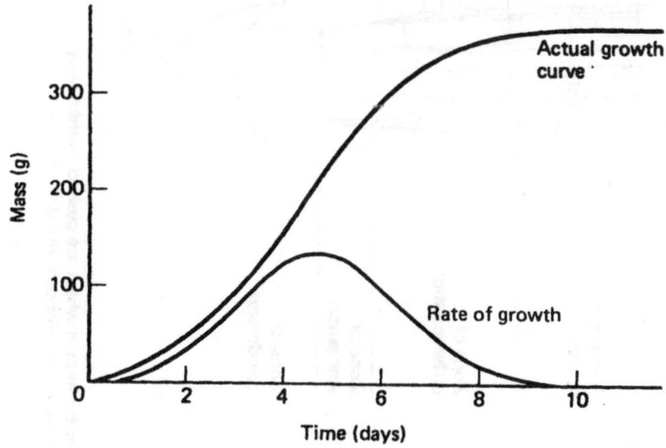

The growth rate curve of most organisms is bell-shaped.

Figure 10.11 (a): Relationship between actual growth and rate of growth.

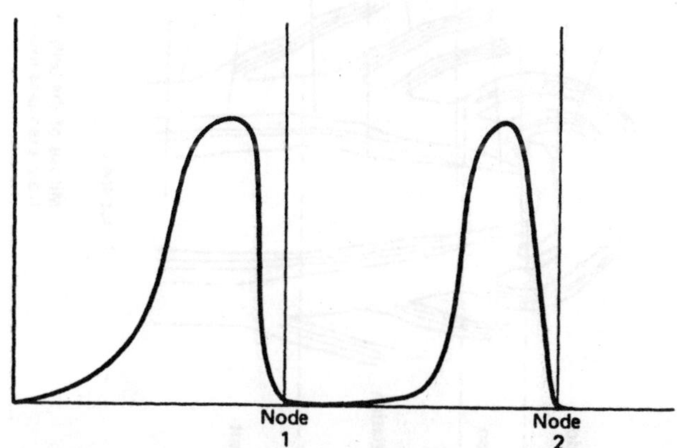

In *Tradescantia* maximum growth occurs at the base of each internode; there is no growth at the nodes.

Figure 10.11 (b): Growth rate of Tradescantia.

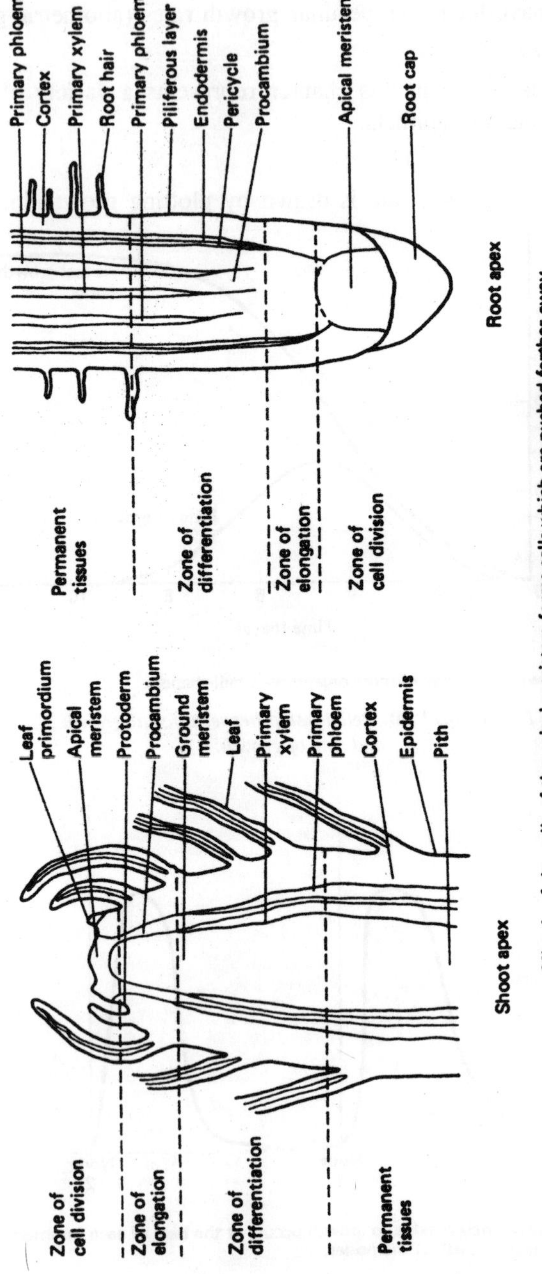

Figure 10.12 (a): Primary (apical) meristem in a shoot.

Figure 10.12 (b): Primary (apical) meristem in a root.

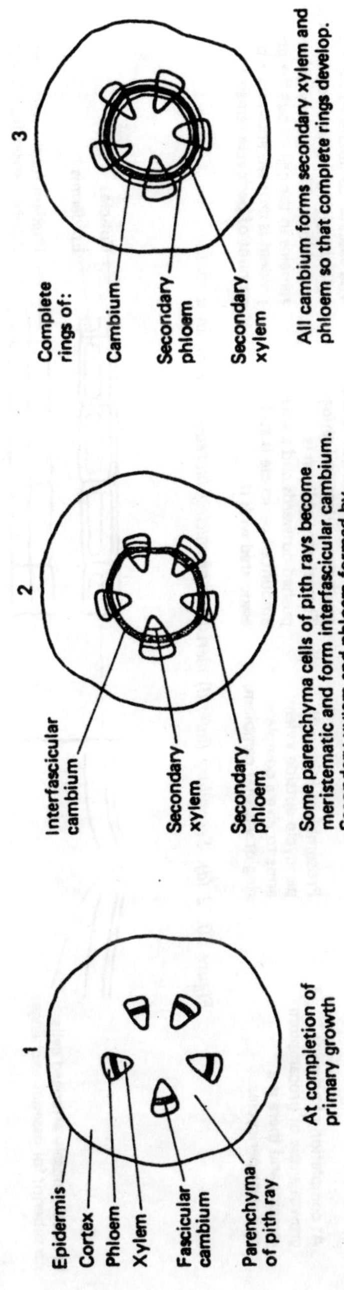

Figure 10.12 (c): Secondary (lateral) meristems producing increae in girth in a shoot.

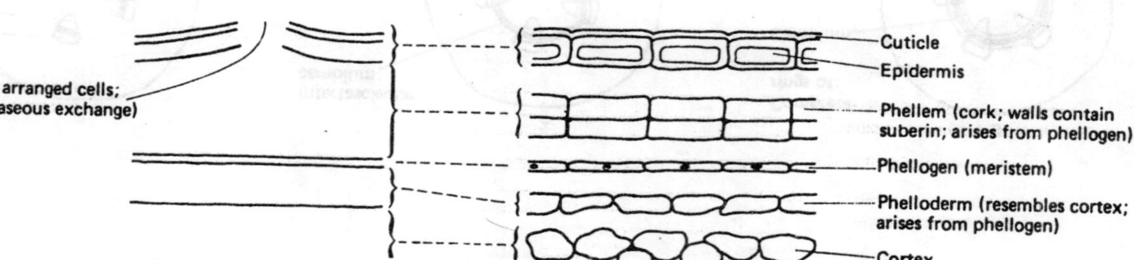

1. At completion of primary growth a row of procambium remains and there is a complete pericycle.

2. Procambium joins with pericycle outside xylem arms to give a complete ring of vascular cambium.

3. The vascular cambium forms secondary xylem and phloem. The primary phloem is pushed outwards and a small amount of pericycle is still associated with it.

4. The vascular cambium forms a smooth cylinder giving rise to secondary xylem and phloem. The primary xylem remains in the centre but the primary phloem is crushed and only a small amount of pericycle remains.

Figure 10.12 (d): Secondary (lateral) meristems producing increase in girth in a root.

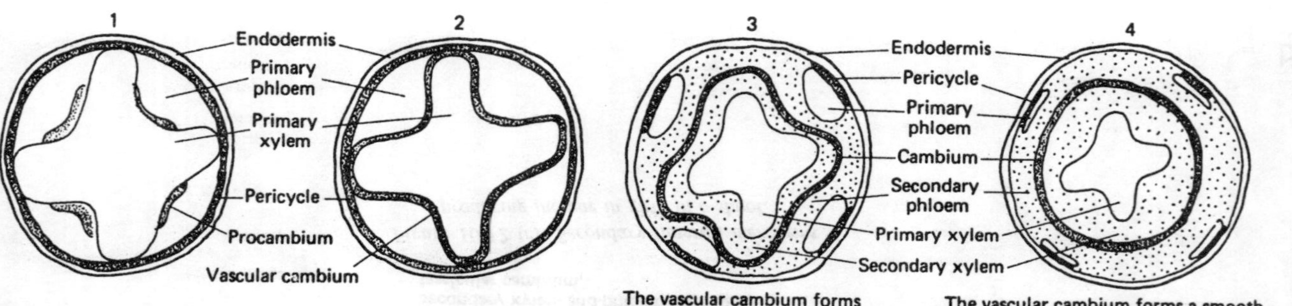

Figure 10.12 (e): Lateral meristems in cork formations.

(i.e. increase in growth over successive periods of time) against time.

The rate of growth in plants and animals is affected by:

1. **Genotype:** dominant and recessive alleles determine such things as protein synthesis, metabolism and size.
2. **Hormones:** auxins and gibberellins in plants and thyroxine and somatotrophin in animals.
3. **Nutrition:** plants require carbon dioxide, light and water; animals need proteins, fats, carbohydrates, vitamins, mineral ions and water.
4. **Environment:** e.g. light for vitamin D synthesis in some animals and for photosynthesis in plants; effect of day-length on meristematic activity; temperature has different effects on ectothermic and endothermic animals; thermoperiodicity in plants influences flowering, etc.

Growth may be limited by such negative factors as disease, pollution and parasites. These are the most important influences on growth but some organisms may be affected by stress, atmospheric pressure and gravity.

Development in Plants

In the early stages of development, cell division occurs throughout the embryo, but as it develops into an independent plant the addition of new cells is restricted to certain parts—*meristems*. The presence of meristems, whose activity permits growth throughout the life of the organism, distinguishes plants from animals.

Meristems may be apical, i.e. located at the tips of main and lateral roots and shoots, or lateral, arranged parallel to the organ in which they occur, e.g. vascular cambium and cork cambium (phellogen).

Typically meristematic cells have large nuclei, compact cytoplasm and small vacuoles. When conditions are favourable these cells divide by mitosis, then elongate and differentiate.

Development in Animals

In contrast to plants, development takes place all over the body in animals. Embryonic development is triggered by fertilization and the multiplication of cells mostly ceases after the organism reaches adult size; the number of organs remains constant.

Cleavage

Following fertilization the nucleus of the zygote divides mitotically, each division being accompanied by cleavage of the cytoplasm to form

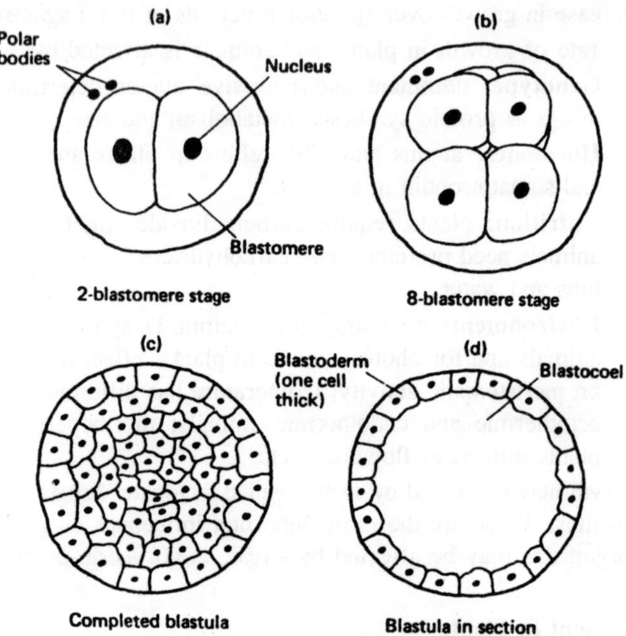

Figure 10.13: Cleavage in Amphioxus.

separate cells, or blastomeres. These divisions give rise to an embryonic structure called a blastula whose form depends on the amount of yolk in the egg.

The description given below and in the diagrams is for *Amphioxus,* which is later distinguished from the frog, fowl and mammal. In *Amphioxus* the egg contains relatively little yolk (i.e. is microlecithal) and the segmentation is holoblastic (cleavage of all the cytoplasm occurs).

The blastomeres are more or less equal in size although those at the vegetal pole may be slightly larger and fewer due to the accumulation of yolk.

Frog

The egg is larger and contains more yolk. Cleavage is holoblastic but more unequal. The cells at the animal pole are smaller and more numerous than the yolky ones at the vegetal pole. The blastocoel is confined to the upper region of the blastula.

Fowl

The very large amount of yolk (macrolecithal) permits only partial

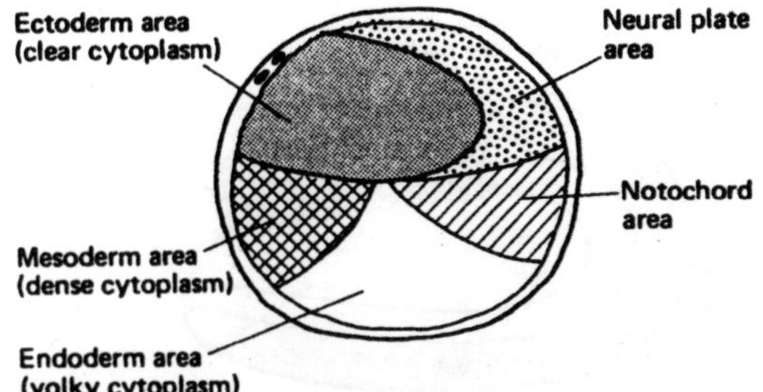

Figure 10.14 (a): Presumptive areas of blastula in Amphioxus.

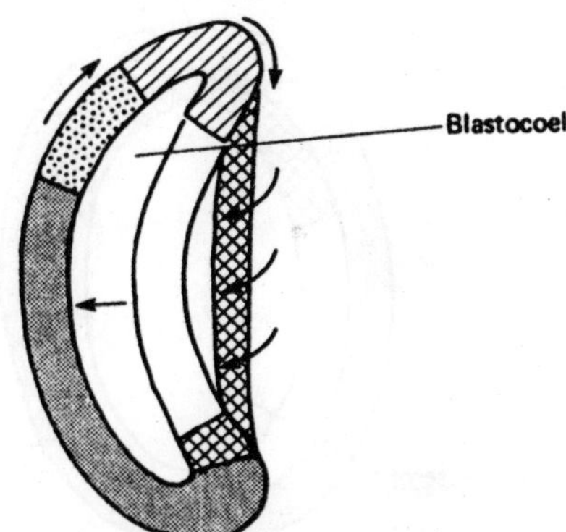

Figure 10.14 (b): Movement of presumptive areas during gastrulation in Amphioxus.

cleavage (i.e. it is meroblastic), which is confined to a small region at the animal pole and results in the formation of a small cap of blastoderm, the blastodisc, and beneath this the sub-germinal cavity.

Mammal

The egg is minute and contains very little yolk (microlecithal). Cleavage is holoblastic but results in a solid ball, or morula, and not in a hollow blastula.

Gastrulation

Gastrulation follows cleavage and results in the formation of a hollow gastrula comprising a wall of two distinct cell layers and a cavity, the archenteron. The blastoderm of chordates consists of definite regions (presumptive areas) destined to give rise to the ectoderm, endoderm, mesoderm, neural plate and notochord.

During gastrulation the cells of these presumptive areas are moved or migrate to their correct positions in the gastrula.

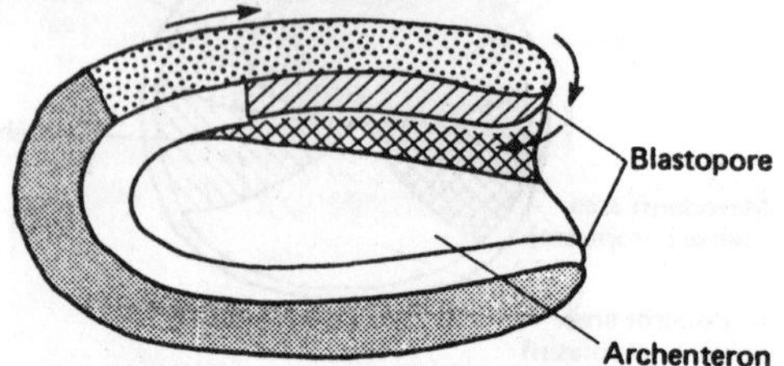

Figure 10.14 (c): Elongation of the gastrula.

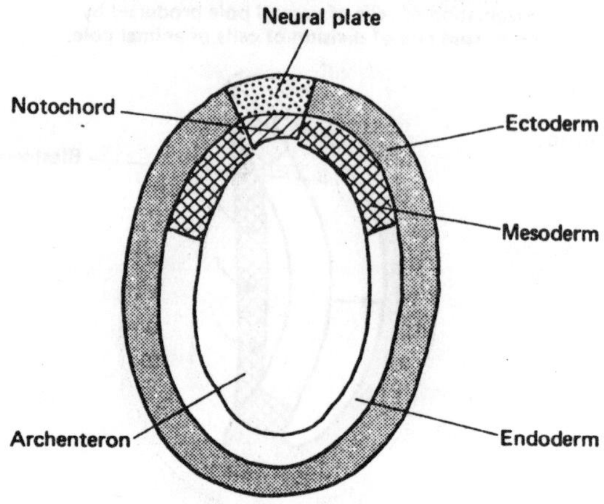

Figure 10.14 (d): TS gastrula.

Gastrulation is modified by the amount of yolk present in the egg.

Frog

The thickness of the blastoderm at the vegetal pole prevents the formation of a gastrula by simple invagination. The rapidly dividing cells of the animal pole migrate over the surface of the yolky cells at the vegetal pole, a process known as overgrowth or epiboly.

At the same time, along a small crescentic area on the dorsal side, cells start to migrate beneath the outer layers. This groove is the beginning of the archenteron.

Fowl

A primitive streak and groove appear on the blastodisc and in this region migration of cells occurs during gastrulation. The sub-germinal cavity becomes the archenteron.

Mammal

The morula divides to form a hollow ball of cells whose wall is one cell thick and whose cavity encloses a mass of cells later known as the germinal disc. The process of gastrulation is similar to that of the fowl.

CONTROL OF GROWTH

Underlying Principles

The size of an organism and its parts is broadly determined genetically, but changing environmental circumstances and the seasonal use of certain organs (e.g. reproductive ones) require variation in growth within these genetic limits.

In addition plants have no contractile tissue to allow them to move, and yet their survival may depend on movement to or from stimuli such as light and water. Such movements can only occur as part of the growth process. It is clear that this variation in growth must be carefully controlled if it is to be of value and the absence of such control may be detrimental (e.g. cancers).

The control is largely exerted by hormones in animals because their effects are more general than the specific responses of the other major controlling system, the nerves. In addition growth is a relatively slow process often involving permanent change.

This is more suited to hormonal control than the rapid and temporary response brought about by nerves. In the absence of a nervous system, the control of growth in plants must be hormonal.

Points of Perspective

It would be an advantage for a candidate to have carried out a number of experiments involving tropic and tactic responses and to have observed the effects of plant hormones on growth.

Auxin concentrations can be measured by collecting the hormone in a gelatin block then placing it on one side of a decapitated coleoptile. The degree of curvature it causes under fixed conditions is a measure of the auxin concentration.

This is an example of the important technique of bioassay, and an understanding of the principles underlying the process would be of value.

A range of practical applications for plant growth hormones is important, e.g. hormone weed killers and defoliants; artificial ripening of fruit by ethene; fruit development in the absence of pollination by auxins.

Likewise a knowledge of the medical application of animal hormones and the effects of overproduction and under-production would be helpful.

The use of growth hormones in factory farming could also be considered. Finally candidates may consider the consequences of uncontrolled growth, in particular the production of tumours (cancers) and the problems associated with their cure.

Essential Information

Control of growth in plants

Plant growth substances, or hormones, are produced in very small quantities in one part of the plant and transported to another part where they promote, inhibit or in some way modify growth.

There are five main classes of plant growth hormones, which also affect other activities of the plant:

1. Auxins
2. Gibberellins
3. Cytokinins
4. Inhibitors
5. Ethene

Phytochromes

A number of plant growth reponses are influenced differently by light of different wavelengths. For light to have an effect it must be absorbed by a photoreceptor substance.

In about 1960 the pigment phytochrome was isolated; this exists in two interconvertible forms: one which absorbs red light with a peak at about 660 nm (P_{660}) and the other which absorbs far-red light with a peak at about 730 nm (P_{730}).

$$P_{660} \underset{\substack{\text{far-red light = rapid conversion} \\ \text{dark=slow conversion}}}{\overset{\text{red light (daylight)}}{\rightleftarrows}} P_{730}$$

Phytochrome is present in the leaves. After absorbing the appropriate wavelength of light, it causes the conversion of a hormone precursor to a hormone which then affects growth.

Photoperiodism

Flowering is regulated by daylength. Three basic groups of plants exist although all intermediates between them may be found:

1. Long-day plants, e.g. clover, barley, radish, petunia. These only flower when the light period in a 24-h cycle exceeds about 10 h. In temperate regions these plants flower in the summer.

2. Short-day plants, e.g. tobacco, cocklebur, poinsettia, chrysanthemum. These only flower when the light period is shorter than about 14 h. In temperate regions they generally flower in the spring or autumn.
3. Day-neutral plants, e.g. carrot, violet, begonia, cucumber. These are indifferent to daylength.

Long-day plants are thought to flower when the presence of red light, or a long period of sunlight, causes a sufficient accumulation of P_{730} and thus a low level of P_{660} Short-day plants flower when far-red light, or a long period of darkness, causes a sufficient accumulation of P_{660} and thus a low level of P_{730}.

Other effects of red and far-red light (through the phytochrome system):

1. Germination: some seeds, e.g. *Lactuca,* germinate better in far-red than red light.
2. Formation of some plant pigments: e.g. red light induces the formation of anthocyanins.
3. Elongation of internodes: in many plants the internodes lengthen in far-red light and elongation is inhibited by red light.
4. Expansion of leaves: leaf area increases in response to red light.

Control of growth in mammals

Hormones play a major role in regulating growth in mammals and they are co-ordinated by the activity of the hypothalamus.

$$\text{Hypothalamus} \xrightarrow[\text{hormones}]{\text{1st order}} \begin{matrix}\text{Pituitary}\\ \text{gland}\end{matrix} \xrightarrow[\text{hormones}]{\text{2nd order}}$$

$$\begin{matrix}\text{Endocrine}\\ \text{glands}\end{matrix} \xrightarrow[\text{hormones}]{\text{3rd order}} \begin{matrix}\text{Widespread}\\ \text{effects}\end{matrix}$$

The main hormones influencing growth in mammals are:

1. Somatotrophin
2. Thyroxine
3. Insulin
4. Sex hormones
5. Cortisol.

Hormonal control of insect metamorphosis

Insect metamorphosis is controlled by three main hormones:

1. Brain hormone—produced by neurosecretory cells in the brain

and stored in the corpus cardiacum.

2. Ecdysone (moulting hormone)—produced by the prothoracic gland when stimulated by brain hormone.
3. Juvenile hormone—produced by the corpus allatum.

All moults require ecdysone. The type of moult is controlled by the concentration of juvenile hormone.

High concentrations of juvenile hormone cause larval moults.

Low concentrations of juvenile hormone cause pupal ecdysis and a pupa is formed. Absence of juvenile hormone causes imaginal ecdysis and an imago is formed.

11

Environmental Approach

ECOLOGY

Underlying Principles

The Law of conservation of energy

Energy may be transformed from one form into another but is neither created nor destroyed.

Biotic factors and abiotic factors are interrelated. Biotic factors which influence an environment include feeding interrelationships, i.e. food chains, food webs, and biomass pyramids. Examples are

1. Grazing
2. Predator-prey balance
3. Parasitism
4. Symbiosis.

The effects of abiotic factors will vary according to the environment. Table elsewhere in this chapter lists some of these.

A knowledge of practical field techniques will be essential for an understanding of ecology. The ability to use keys to identify the organisms in the environment studied is a vital part of ecology.

Essential Information

1. **Biosphere:** the sum total of the living organisms on Earth and their environment.
2. **Biome:** an aggregation of similar ecosystems in a particular

Table 11.1: Abiotic factors.

Physical	*Chemical*	*Edaphic*
Temperature	Oxygen	Chemical and physical factors related to soil type and important in terrestrial environments
Light	pH	
Wave action (marine)	Minerals	
Desiccation (littoral)		
Substrate		
Points of perspective		

region of Earth. It could be a habitat or zone such as grassland, desert or ocean.

3. **Ecosystem:** a localized group of communities and their physical environment.
4. **Community:** a localized group of several populations of different species.
5. **Population:** a geographically localized group of individuals of the same species.

One way of investigating biotic factors which affect any environment is to classify the various organisms according to their methods of nutrition. The basic distinction is between autotrophs and heterotrophs.

1. Autotrophs: organisms able to manufacture organic foods from inorganic nutrients.
2. Heterotrophs: organisms unable to manufacture food and dependent on autotrophs for food.

The heterotrophs can be sub-divided according to the nature of their food.

1. Herbivores: feeding on plants.
2. Omnivores: feeding on plants and animals.
3. Carnivores: feeding on animals.

In addition to the predatory carnivores, the grazing herbivores, and the omnivores that do both, there is another group of heterotrophs which obtain food from living organisms. These are the parasites, which live either in or on their hosts.

Any animal or plant may act as a host to a number of individuals of one or more parasitic species. These, in turn, may be parasitized themselves. Organism's which are parasitic on parasites are called hyperparasites. In this way, special category food chains may be set up:

Host → Parasite → Hyperparasite

Among the organisms which feed on dead organisms, there are two major types:

1. Those that take in their food as solid masses which are digested in the gut. These are the scavengers or, when the food particles are very small, detritus feeders (detritivores).
2. Those that secrete digestive enzymes onto dead organisms or their wastes and absorb the products of digestion as nutrients. Examples are bacteria and fungi. They are called saprophytes.

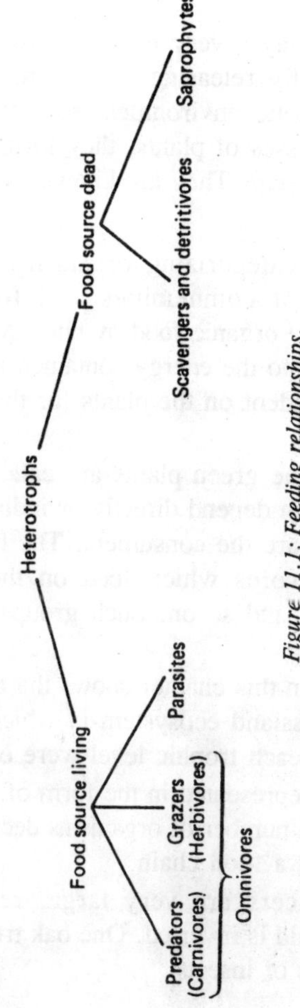

Figure 11.1: Feeding relationships.

There are also parasitic fungi and bacteria.

In addition to breaking down the organic chemicals in the dead bodies of other organisms or their wastes, the digestive activity of saprophytes also produces inorganic materials such as water and minerals, which return to the abiotic environment. This results in putrefaction or decay.

Living organisms occupy a position in food chains as herbivores, carnivores or as producers. These chains, which always have plants as their starting point, are often called grazing chains. When an organism dies, it can act as a starting point of another series of food chains containing scavengers, detritus feeders and saprophytes. These chains are decay chains.

The saprophytes play a very important role in the circulation of energy and nutrients. By releasing energy from dead organisms and returning it to the abiotic environment so that it becomes available for the synthetic processes of plants, they form a vital community of organisms in an ecosystem. They are known as decomposers.

Trophic Levels

Another way of categorizing organisms within an ecosystem emphasizes the role of communities. All food chains start with autotrophs, which make organic food by photosynthesis, converting the energy from sunlight into the energy contained in food molecules. The heterotrophs are dependent on the plants for their supplies of energy-giving food molecules.

For this reason, the green plants are called the producers. Next come the animals which depend directly or indirectly on the producers for their food. These are the consumers. The herbivores are primary consumers, the carnivores which feed on the herbivores, are the secondary consumers, and so on. Such groups are known as trophic levels.

Table elsewhere in this chapter shows the results of a quantitative investigation of a grassland ecosystem in which the total numbers of organisms occupying each trophic level were estimated.

The data can be represented in the form of a pyramid of numbers. This illustrates that the number of organisms decreases from one trophic level to another along a food chain.

When the producers are very large, relative to the primary consumers, the pyramid is inverted. One oak tree can provide food for hundreds of thousands of insects.

Table 11.2: Numbers in the trophic levels of a grassland ecosystem.

Trophic level	*Numbers of individuals/m²*
Producers	3100
Primary consumers	400
Secondary consumers	150
Tertiary consumers	0.02

Energy relationships of trophic levels

Energy enters the trophic levels by photosynthesis. It is used for the synthesis of new plant material in growth, repair and reproduction. Ecologists say that it is used to increase the biomass of the trophic

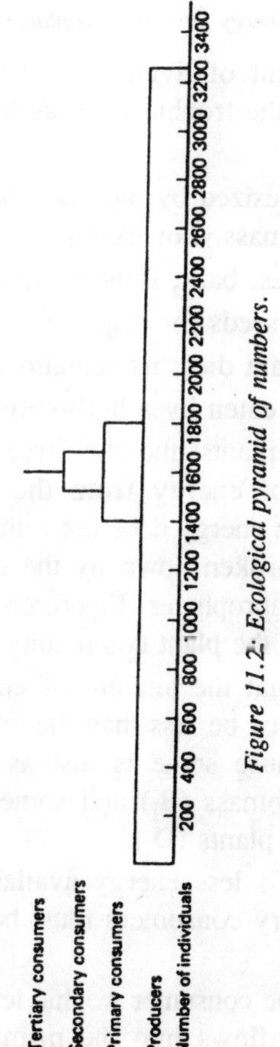

Figure 11.2 Ecological pyramid of numbers.

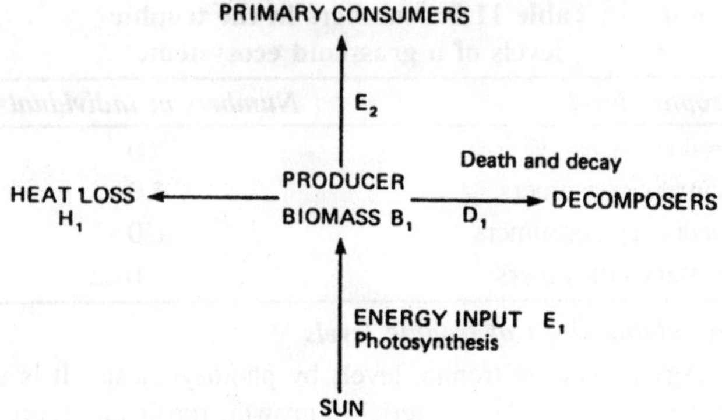

Figure 11.3 (a): Energy flow through plant community.

level. *Biomass* is the weight of living material per unit volume or area. Energy flows out of the trophic level as follows:

1. Heat loss
2. In structures synthesized by the plant but not contributing to an increase in biomass. For example:
   (a) Shedding leaves, bark, flowers, fruits.
   (b) Production of seeds for dispersal.
3. *Death*. When a plant dies, its remains contain some energy.
4. The plant may be eaten by a herbivore.

The remains of dead plants and the structures mentioned in 2 above, represent a loss of energy from the trophic level to the environment. However, this energy does not remain locked up in these siructures. They may be broken down by the decomposer chain, so that the energy is used by saprophytes. Figure elsewhere in this chapter shows energy flow through the plant community.

It follows from this that the amount of energy available to the primary consumers (E_2) must be less than the amount originally fixed by the producers (E_1)because some is lost as heat (H ,), some is present in the producer biomass (B_1) and some occurs as non-living structures produced by the plants (D_1).

As a result, as there is less energy available for synthesis, the total biomass of the primary consumers must be smaller than that of the producers.

Energy flow through the consumer trophic levels can be considered in the same way. Energy flows into the primary consumer trophic

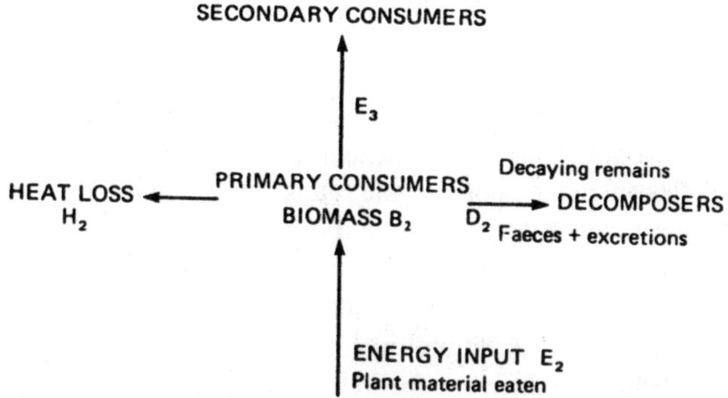

Figure 11.3 (b): Energy flow through primary consumer trophic level.

level in the form of plant material eaten (E_2). Some of the energy contributes to the primary consumer biomass (B_2), some is lost to the environment as heat (H_2) and some passes to the decomposers (D_2) in the form of dead bodies, undigested material in faeces and excretory products. Finally, some passes on to the secondary consumer level as herbivores eaten by carnivores. Figure elsewhrere in this chpater summarizes the energy flow through the primary consumer trophic level.

In this way, a picture of energy flow through each trophic level and ultimately the whole ecosystem may be built up.

Ecological succession

After a patch of ground has been cleared of vegetation, certain plants begin to establish a simple community. Gradually other species spread into the community where they often displace some of the original colonizers. It will be noticed that the changes

take place rapidly at first, slowing gradually until dynamic equilibrium is reached and the community is stabilized. The sequence of change is called ecological succession and is characteristic of all ecosystems.

The first to become established are the *pioneers*: the equilibrium community at the end of the succession is the *climax*. During the succession, many environmental factors will change.

Some change as a result of the influence of the community on the abiotic factors of the ecosystem; others may be the result of import or export of materials to or from the ecosystem. As the abiotic factors change, the community changes in response and succession proceeds under feedback control as shown in Figure elsewhrere in this chapter.

Progressive changes in communities during ecological succession can be illustrated with the model shown in Figure elsewhere in this chapter.

Communities Individual populations interact with one another as food supplies or as population regulators, e.g. a prey and its predators. However, populations also interact with other members of their communities in much more subtle ways which may have little direct effect in the regulations of the specific groups involved, but which are fundamental to the maintenance of the community as a whole, and hence to the existence of any species in the community.

For example, a rabbit and nitrogen-fixing bacteria in a grassland ecosystem belong to different food chains and are regulated by different components of the abiotic and biotic environment; yet without nitrogen-fixing bacteria, nitrogen could not be cycled and rabbits could not exist. Interdependence among species goes far beyond the obvious interactions between one population and another. Populations exist, in fact, because they are parts of communities.

The effects of climatic, edaphic and biotic factors on distribution

All organisms must maintain themselves in an external environment which could harm them because it differs from their internal environment. However, when the external environment is variable, then the problem is greater than it would be if the environment remained stable. Table elsewhere in this chapter shows how the marine environment is least variable and the terrestrial environment most variable.

Distribution of organisms within the environments listed in Table elsewhere in this chapter will depend on the physical and chemical factors indicated in the table.

The edaphic (soil) and climatic factors will determine the presence or absence of various types of vegetation. Because carnivores eat herbivores, and herbivores eat plants, the distribution of plant species governs the distribution of animal species.

The basic nutritional requirements of plants are carbon dioxide, water and minerals, so their distribution might be expected to be either uniform or random. Rainfall, light intensity, daylength and temperature are climatic factors which determine the pattern of the world's vegetation.

Vegetational zones are named after their climax vegetation—the natural vegetation which would grow if there were no human interference. The distribution of organisms therefore depends on:

1. Whether the species reaches the area
2. Suitable climatic and edaphic factors
3. An appropriate food source
4. Whether the species can survive competition or predation.

The ecological significance of soil

The soil does not offer an inexhaustible supply of mineral ions to plants. Many essential elements may be in short supply within the community.

During the millions of years of evolution, the soil and atmosphere would have become completely emptied of elements if it were not for recycling processes. The micro organisms which use soil as a habitat are extremely important in recycling many elements, for example carbon and nitrogen.

The Carbon Cycle

Green plants absorb carbon dioxide from the atmosphere and utilize if for the manufacture of carbohydrates, proteins and fats. The plants are eaten by herbivores which digest and assimilate the foods originally synthesized by the plants. Carnivores continue the food chain, but eventually death ensures that all links of the chain are returned to the soil.

No matter how long or complex the food chain/web, the carbon which was originally incorporated into the producers sooner or later forms part of dead organisms upon which saprophytes feed, or is returned to the atmosphere as carbon dioxide from respiration. The carbon in the dead organic matter becomes converted to carbon dioxide by putrefying bacteria and fungi. As a result of these processes, the concentration of carbon dioxide in the atmosphere remains fairly constant.

The Nitrogen Cycle

Soil-dwelling bacteria are some of the few organisms which can utilize atmospheric nitrogen. They are able to convert it to nitrates which can be absorbed by plants and used during protein synthesis. Herbivorous animals acquire their nitrogen from plants, whereas carnivores rely on animal protein for their nitrogen supply. The removal of nitrate from the soil is balanced by its return as a result of bacterial activity.

The circulation of nitrogen within the community is similar to that of carbon. Nitrogen is passed from the plants to herbivores and along a food chain. Saprophytes eventually decompose the tissues of all the organisms concerned and so liberate the nitrogen once more. They liberate it as ammonia which reacts with carbon dioxide and water present in the soil spaces. At this point in the cycle, two species of bacteria present in all soils play a vital part. One, called *Nitrosomonas, is* able to derive all the energy it needs from that released when ammonium carbonate is oxidized.

$$(NH_4)_2CO_3 + 3O_2 \rightarrow 2HNO_2 + CO_2 + 3H_2O + ATP$$

The nitrous acid reacts with salts in the soil to produce nitrites. These are acted upon by *Nitrobacter* which derives its energy by oxidizing nitrites to nitrates.

Another method of nitrate production is via a species of bacterium called *Rhizobium leguminosarum,* which lives in swellings on the roots of plants belonging to the family Leguminosae. They can incorporate atmospheric nitrogen into their protoplasm.

The host plants absorb some nitrogenous molecules from the bacteria. Furthermore, the whole community eventually benefits from their activity because, when the host plant dies and is decomposed, a greater amount of nitrogen is released in the soil than would otherwise be the case.

MAN AND HIS ENVIRONMENT

Underlying Principles

Primitive man affected his environment by hunting, fishing and removing trees for fires and shelters. He was nomadic and therefore any effect he had on an area was temporary and the environment had adequate time to recover. About 11000 years ago man began to cultivate his own crops and thus communities settled in one place in order to harvest and store the crops.

The use of tools and later domestication of animals led to more efficient agriculture, the ability to produce more food and hence support a larger population. Demands were created on the environment and trees were felled to provide shelters and more land for cultivation. The cultivation of crops was aided by domesticated animals. The energy therefore came from the food they ate and this limited its efficiency.

With the advent of fossil fuels all this changed and little of the energy in the crops grown went back into growing the crops. Instead machines driven by fuels helped cultivate the crops, the majority of which.wdre available to support a larger population which in turn required more efficient agriculture to support it.

Fertilizers, pesticides and rapid transportation all developed and brought with them pollution problems in addition to those caused by burning the fossil fuels. Such fuels are a finite source of energy and hence there was pressure to develop new energy sources. Nuclear energy was developed with its consequent environmental problems.

Points of Perspective

Answers to questions on this topic are all too often vague and lacking in specific detail. High marks will be obtained by a candidate who can give:

1. The precise names of pollutants
2. Figures concerning lethal doses or the quantities of pollutants produced
3. The exact nature of the effect of a pollutant
4. A wide range of examples.

It would be of benefit to have some knowledge of the effects of pre-industrial man on the environment. A good candidate will understand the principles and methods of conservation.

Essential Information

Man's effects on the environment are largely a result of two factors:

1. The need to produce human food supplies
2. The population explosion.

Human Food Supplies

Agriculture

Probably the first attempts at cultivation began with the deliberate use of fire to clear large areas of land for planting; by doing so, man began to alter the structure of communities. Climax vegetation was destroyed and replaced by monoculture, e.g. a field of cereal crops replaced a forest of oak.

It was soon discovered that if the same crop was grown for several successive seasons, the gross yield fell because crops have specific demands on the soil's resources. The problem could be avoided by crop rotation or by replacing lost materials with fertilizers. The latter solution raises another problem because excess use of fertilizers often causes pollution of rivers due to leaching and draining.

Ecologically, agriculture is contrary to the natural development of communities. Monoculture results in large areas planted with a single species of plant instead of the climax Zommunity of many species. As a result, weeds and pests compete with the single species for environmental resources. When man attacks the competitors with herbicides and pesticides, he creates additional problems of pollution of soil and water by toxic chemicals.

Agriculture provides enormous food resources for those herbivores which share the same food as man. With many plants of one species grown close together, there is no problem of individuals finding new food plants at the dispersal stage.

Thus, large numbers of herbivores can build up at the expense of crops. Whole ecosystems are therefore affected by the change in the original climax community.

The population explosion

There has been a rapid increase in human population since the seventeenth century and this has inevitably had its effect on the environment. Ancient agricultural races have been replaced in certain parts of the world by populations founded on the industrial and technological revolutions.

The results have been an increase in the use of non-replaceable resources, e.g. fossil fuels, and an increase in pollution of land, sea and air by products of fuel combustion.

Solutions to these problems rely on co-operation between ecologists, politicians and economists. A costeffective method of reducing pollution is usually not easy to find and conservationists must argue their case against industry and other cost-conscious sections of the community.

There is no dispute about the scientific aspects of the ecological situations involved. An increase in population imposes greater strains on food supply and its distribution. There are basically two contrasting points of view relating to the food supply problem.

1. Many agriculturalists believe that, with the use of fertilizers, herbicides, and pesticides, the world can produce enough food for the predicted growth rate of population.
2. In contrast, many ecologists disagree and argue that, already at least half the world's population are under-nourished and they suggest that it will be impossible to feed adequately the probable population forecast for the year 2000.

There are three main methods of improving this situation:

1. To stabilize the human population.
2. To increase the efficiency of food production and utilization.
3. To develop new food sources.

At present we lose a great deal of food during its transport from the source of production to the consumer. This is due to competition between man and other organisms, e.g. rats, mice, insects and moulds. Improvement in techniques for combating these is feasible.

Some food is lost for economic reasons because of high transport costs from regions of overproduction to regions of need. There are instances of couptries burning surplus food or allowing it to rot while third world countries have populations suffering from malnutrition and starvation.

It is very easy to confuse biological and political problems in a discussion concerning human food supplies. The biological problems relate to the efficiency with which different types of food can be produced in different environments.

Since energy is lost at each link of the food chain, the most efficient way for an omnivore like man to tap solar energy is to eat plants. The highest primary production comes from tropical crops, e.g. sugar-cane which can grow all the year round but most of the plant's body is not eaten by man. Usually only parts, such as fruit, seeds, leaves or roots are used so that much primary production is not used.

Plant proteins differ from animal proteins in the proportions of amino acids. Herbivores have to eat large quantities of food to obtain certain essential amino acids. Eating animals is economical in terms of the bulk of food consumed although it is wasteful of solar energy. Of the present supply of protein for human food 70% comes from vegetation and 30% from animals. Intensive livestock rearing is the most efficient way of producing those animal proteins which are traditional foods.

The mass production of chicken and veal relies on processed diets which incorporate parts of organisms normally discarded as inedible. Occasionally antibiotics and hormones are added to the diets with subsequent dangerous side effects to the secondary consumers.

Ideally, protein should form about 10% of the total food intake in terms of energy; this constitutes the main difference between diets of well-fed and under-fed races of the world. An adequate diet should contain about 44 g per day protein. In the USA, the average consumption is about 64 g per day, whereas some third world countries have people surviving on 9 g per day.

In 1962, there were 0.5 hectares of arable land per person and it is estimated that the figure will be 0.3 hectares per person by 2000 AD. This implies that for agriculturalists merely to maintain the present levels of nutrition would mean doubling the productivity of arable land. It is possible that marine production could be doubled and terrestrial protein production increased by about 50%.

Other possible sources of food could depend on 4dvances in biotechnology. Certain bacteria can be used to convert petroleum into protein and this so-called single-cell protein production could be an answer to the world's food shortage.

However, these sources can only be exploited through energy-consuming industrial processes and the amount of energy needed to produce edible protein may be so great that the source is totally uneconomic. Nuclear energy or the use of solar energy will probably be essential if a large human population is to survive in the future, despite the potential harmful effects of the disposal of nuclear waste.

Pollution

Pollution is a difficult term to define. It is derived from the Latin word *polluere* which means 'to contaminate any feature of the environment'. It may be broadly said to be 'adding to the environment a potentially hazardous substance or source of energy at a rate faster than the environment can accommodate it'. In this broad sense the definition includes not only man's activities but also certain natural processes.

Air pollution

This form of pollution is largely a result of burning fossil fuels such as coal, coke, or fuel oil. Smoke contains carbon particles, carbon dioxide, sulphur dioxide and fluoride. Domestic combustion is of importance in those areas which have not introduced smokeless fuels. Smoke is a major health hazard as it affects the respiratory tract, increasing the incidence of bronchitis, and may even be a carcinogen.

Normally, the cells lining the bronchial tubes produce mucus to moisten the surface and trap bacteria. When cells are diseased, they cannot function properly and are susceptible to bacteria. Whenever fog builds up in air heavily polluted with smoke, a condition called smog is produced. The dampness, combined with the effects of suspended carbon in smoke, increases the incidence of pulmonary disease.

The problem of reducing industrial smoke is complex; it has been tackled in a number of ways. In 1956, the British government introduced the Clean Air Act which made it illegal to discharge smoke in many areas. This meant that households and factories had to use smokeless fuels, e.g. electricity, gas, oil or solid smokeless coals. The introduction of electric, diesel, and gas turbine locomotives also reduced air pollution from the level it was when steam engines were in use. Modern technology allows fuels to be burnt more efficiently, and washing processes can be installed to remove soluble and heavy particles before the gases are sucked up the chimney where sieves or electrostatic precipitators are also installed.

To allow efficient combustion of petrol, lead 'anti-knock' agents are added and these lead compounds are released into the atmosphere along with carbon monoxide and various nitrogen oxides. The lead is easily absorbed through the lungs and may cause mental retardation in children. Non-lead 'anti-knock' agents are available but are more expensive.

When fossil fuels are burned sulphur dioxide is discharged into the atmosphere. Coal and coke contain about 1-2% sulphur. Much of this can be removed but the cost is high. Control of this form of pollution depends on high chimneys to carry the fumes away. In some areas, problems have arisen when it was washed out from the waste gases by rain as sulphuric acid. This corrodes buildings. Plants are particularly vulnerable to sulphur dioxide pollution because the gas is absorbed through stomata and is absorbed into the leaf cells in lethal doses.

Fluoride is a waste-product of certain industrial processes involved in the manufacture of pottery, bricks, steel and aluminium. It is absorbed by plants and becomes concentrated in leaves. Cattle and sheep grazing on the plants become affected over a period of time. Bones become soft and joints stiffen. Methods used to get rid of fluoride depend on building higher chimneys. Other methods have proved very expensive.

Aquatic pollution

Freshwater pollution

Man soon realized that rivers are the easiest and cheapest means of transport. Towns grew alongside them, and the simplest way to get rid of unwanted material, including sewage, was to put it into the river. Rivers became the first sewers and, before the population explosion of the industrialized communities, were efficient as such and still remained healthy.

However, when the bulk of the waste became too much, bacteria thrived and began to deplete the oxygen from the rivers causing a *biological oxygen demand (BOD)*. In still waters the effects have been marked with many lakes in the world 'dying' as a result. Mass epidemics have spread via water systems because of pollution by pathogens. It was not until the 1800s that sewage works were constructed to treat the waste before discharge.

Toxic wastes from industrial sources are numerous and varied. The main danger is from organic chemical and fertilizer manufacturers. Heavy metals act as non-competitive inhibitors of enzymes and build up via food chains once they enter the environment.

Pesticides, herbicides and fungicides drain away from fields, enter waterways, and are concentrated through food chains so that top carnivores become poisoned. Excessive use of fertilizers can also give rise to a build-up of toxic by-products by leaching and draining.

Thermal pollution becomes a problem where large quantities of heated water are discharged, e.g. from power stations. The consequent rise in temperature not only kills organisms directly, but reduces the solubility of oxygen in water, thus killing aerobic organisms.

Anaerobes can thrive in these conditions and compete successfully with other forms of life for environmental resources. Detergents are rich in phosphates and these provide nutrients for the vegetation in rivers and lakes. Toxic by-products from algal blooms result in the death of many organisms

The prevention of these forms of pollution depends on treating waste-products at the source so as to render them harmless. Cost seems to be the major problem because the expense of treatment must be met by increasing the cost of the end-product manufactured or by reduction of profit margins.

Marine pollution

Oil pollution of the sea is one of the most emotive forms of pollution probably because its effects are direct and easily visible. Illegal washing of oil tanks at sea, together with accidental loss of oil by collision, wrecks, and oil rigs provide the biosphere with one of the most unsightly forms of pollution.

The oil directly affects vertebrates which come in contact with it, e.g. sea birds, seals and fish, with the physical effect of preventing gaseous exchange. Planktonic invertebrates are also poisoned and so whole food chains become affected.

Methods of treating oil pollution have had limited success. They include absorbent methods involving covering oil with an absorbent material such as straw and collecting the oil-soaked material. Another approach is to spray the oil with a solution of plastic. The plastic hardens to form a coating over the oil. Plastic foam and sawdust have also been used but the problems with these methods include cost effectiveness and the difficulty of spreading the absorber evenly and then collecting it at sea.

Alternatively, heavy materials may be added to the oil causing it to sink. Bacteria break down the oil quite rapidly when it is on the sea bed. Sand, treated chemically to make oil cling to the grains, has been used to sink oil and because very small amounts of chemicals are needed to treat the sand, it is cheap and safe to marine life.

Calcium sulphate and pulverized fuel ash, when treated with silicone, are other effective materials for sinking oil. Detergents have been used to reduce the surface tension of oil but unfortunately may damage wildlife when ingested.

Radioactive pollution

One of the unavoidable consequences of the use of nuclear energy is the production of radioactive waste. Highly reactive, long-lived wastes are usually concentrated, sealed into lead containers, and dumped in the deep sea. Wastes with a lower level of activity are dispersed by the sea, which dilutes them to acceptably safe levels. The amounts which can be discharged are strictly controlled.

One of the most dangerous effects of radiation is the damage it causes to human chromosomes. Mutations often result. These are often passed from one generation to the next before they exert their often damaging effects.

In considering the dangers posed by a radioactive substance a number of factors need to be taken into account. These include the half-life of the substance, its possible concentration in food chains and its role in an organism. ^{90}Strontium for instance is particularly hazardous to humans:

1. It is produced by nuclear explosions and is therefore constantly produced.
2. It has a half-life of 28 years and therefore persists.
3. It is readily absorbed by grasses and concentrated in cows' milk which is consumed in large quantities by humans, especially babies who are particularly vulnerable because their cells are rapidly dividing.
4. It becomes concentrated in human bones, which contain the bone marrow which is rapidly dividing to produce blood cells. This division may therefore become disturbed and leukaemia can result.

A radioactive substance released at a relatively harmless level may become accumulated to dangerously high levels along a food chain. The levels of the phosphorous isotope ^{32}P in a North American river were found to be:

1. In the water - 1 (arbitrary unit)
2. In phytoplankton - 1000
3. In aquatic insects - 500
4. In fish -10
5. In ducks - 7500
6. In duck eggs-200 000 (shells contain calcium phosphate)

Here again the main danger is that the highest levels are found next to the developing embryo which comprises rapidly dividing cells.

Terrestrial pollution

There are two distinct types of terrestrial pollution: waste materials that are dumped and pesticides. Waste heaps, derelict mines and buildings present the following problems:

1. They are unsightly.
2. They waste possibly useful land.
3. They may create a danger to life, e.g. children trapped in mines, the Aberfan slag heap disaster of 1966.
4. They may be a health hazard by attracting vermin.
5. They cause urban decline, causing people to move away which creates social decay.

Pesticides are numerous. It is essential that they are properly used. Often however they 'drift' from their targets or are leached into streams, rivers and lakes. A good pesticide should:

1. be specific only to the pest it is directed at
2. be rapidly broken down and not persist in the natural environment 3 not be accumulated through food chains.

12

POPULATION

Among the more spectacular members of the rich mammalian fauna that occupy the African continent are two large ungulates, the roan antelope (*Hippotragus equinus*) and the sable antelope (*H. niger*). The roan antelope inhabits open or lightly wooded land, whereas the sable antelope travels among the acacias (*Acacia spp.*) of the savanna. In recent decades, sable and roan antelopes have been declining in the southern portions of their range.

In South Africa, attempts are being made to preserve and assist in the recovery of these two species by designating large tracts as nature reserves. Because sable and roan antelopes were locally extirpated, animals were live-trapped in the wild, transported, and released on the nature reserves.

Growth of their populations in many of the 2800-13,000-ha reserves has been slow. In recent years, some herds have declined drastically. When wildlife biologists David Wilson and Stanley Hirst (1977) were asked to determine why numbers of these two antelopes declined on the South African reserves over the past few decades, they were presented with a basic problem in population ecology.

Their task was to identify specific causes for the decline of roan and sable antelopes and to make practical recommendations for their preservation and management. Observers had speculated that such factors as habitat deterioration, encroachment by agriculture, illegal or uncontrolled hunting, and (among roan antelope) the disease called anthrax were responsible for the decline. But no one really knew.

The methods used by Wilson and Hirst as they began to unravel the mysteries of the disappearing antelopes were not unique to studies of African mammals. Similar methods are employed by biologists who work with various animals, whether scarce or abundant, whether they are caribou (*Rangifer tarandus*) in Alaska, cottontail (*Sylvilagus spp.*) in Virginia, or bobwhite (*Colinus virginianus*) in Illinois.

These are problems of population management that encompass basic concepts of population ecology. An extensive body of literature exists on the theory of population ecology, much of which is still evolving. Ecologists attempt to explain such phenomena as population cycles and general causes for the regulation of numbers of animals.

Most of the theoretical population literature is beyond the scope of this book. The purpose of this chapter is to present basic concepts of population ecology and to describe approaches that are useful to biologists who must solve practical problems involving the management of wildlife populations.

SOME DEFINITIONS

A *population* is defined as a group of organisms, usually of the same species, occupying a defined area during a specific time. Populations have characteristics not possessed by individual animals. For example, a population has *density*, meaning a certain number of individuals per unit area: 20 blue grouse (*Dendragapus obscurus*) per 100 ha, or 144 sugar maples (*Acer saccharum*) per ha.

A population has a *birth rate*, or *natality*, defined as the number of births per thousand, per hundred, or per individual per year, and a *death rate*, or *mortality*, defined as the number of deaths per number of individuals per year. A population also has an *age structure*—that is, a distribution of numbers of individuals of various ages.

Naturally, the proportion of individuals of breeding age in a population affects the birth rate and strongly influences growth. Likewise, the proportion of old animals affects the death rate. Populations also have *sex ratios* that influence the reproductive potential.

Fecundity refers to the number of eggs produced per female, or to the number of sperm produced per male. Because sperm numbers rarely influence the birth rate, fecundity nearly always pertains to the number of eggs produced. *Fertility* is the percentage of eggs that are fertile.

Production is the actual number of offspring produced, whether born or hatched, by a population during a specific period of time. Some authorities, however, are more conservative and measure production only as the number of new individuals reaching breeding age (the process is also called *recruitment* when it includes immigration); animals dying between birth and sexual maturity are not counted as recruits.

Changes in such parameters as sex ratio and age distribution greatly influence production.

That is, the number of offspring can vary a great deal depending on the proportion of females and animals of breeding age in a population at a given time.

THE LOGISTIC EQUATION

It has been recognized for some time that animals tend to give birth to many more individuals than will survive to breeding age. If deaths did not offset births, the result would be an infinitely growing population. Under ideal conditions, a population for a time can show a rate of growth that is exponential—that is, it grows at an ever-increasing rate. Such conditions may occur when a small population is introduced into a new and favourable environment.

No shortage of food, cover, or space exists, and no disease, parasites, or predators affect any individuals. The birth rate is maximum, limited only by the reproductive physiology of the species; and the death rate is minimum, with deaths occurring only from old age.

Such conditions have been created in the laboratory for yeast cultures and mouse populations that were provided with room to grow and plenty of food. The equation for such growth is conventionally expressed as:

$$\frac{\Delta N}{\Delta t} = rN$$

where

ΔN = change in number

Δt = change in time

r = the "per head" maximum potential growth rate

N = number of individuals in a population

As an example, suppose we have a population of 50 individuals (N) and each individual has the average capability of contributing one fourth (0.25) of an individual to the population in a given unit of time (r). The change in number per unit time ($\Delta N/\Delta t$) would be expressed as:

$$\frac{\Delta N}{\Delta t} = rN \tag{1}$$

$$\frac{\Delta N}{\Delta t} = 0.25(50) = 12.5$$

The answer, 12.5, is the number of individuals that is added to the population in the time interval (t). This number then must be added to the original population (N_t) to obtain the number in the new population (N_{t+1}), so that our new population is

$$N_{t+1} = N_t + \frac{\Delta N}{\Delta t}$$
$$N_{t+1} = 50 + 12.5 = 62.5 \quad (2)$$

For the next time step, we simply repeat the process using the same r value (0.25) but a new N (62.5), to calculate the number added to the population:

$$\frac{\Delta N_{t+1}}{\Delta t} = 0.25(62.5) - 15.5$$

$$N_{t+2} = N_{t+1} + \frac{\Delta N_{t+1}}{\Delta t} \quad (3)$$

$$N_{t+2} = 78$$

As one can see by continuing this process, the population will grow at an ever-increasing rate. The integrated form of the equation is: $N_t = N_o^{e^{rt}}$, where N_t represents the population at t time intervals, N_0 is the original population, r is the per head potential growth rate, e *is* the base of natural logarithms, and t is the time interval by which r is expressed.

Such an equation is most useful for populations of organisms such as bacteria, yeast, some insects, and possibly some small mammals in which breeding and growth are continuous. For most wildlife populations, which have a distinct breeding season, growth takes place in steps,

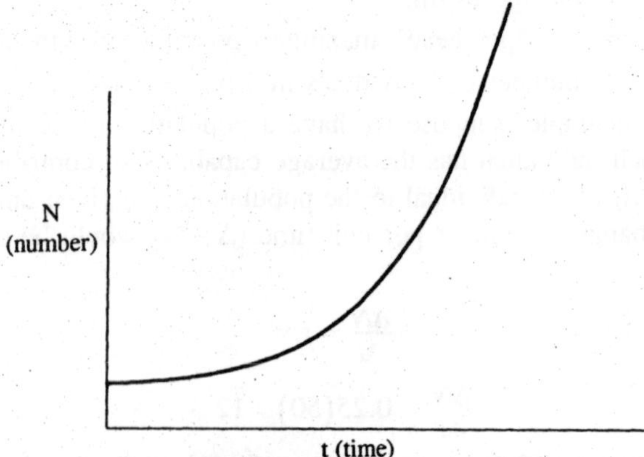

Figure 12.1: Growth of a population with unlimited food and space, $\Delta N/\Delta t = rN$.

and equations (1) and (2) are more appropriately used. Under what conditions might we expect a wild population to express exponential growth? The happy circumstances of practically unlimited food and no biological enemies of whatever size occur very rarely in nature.

But there have been a few instances that have come close to such conditions. These circumstances invariably occur when a population is introduced into a new and favourable environment that has been unoccupied previously by its species. Of the several cases reviewed by Dasmann, only one population, that of white-tailed deer (*Odocoileus virginianus*) introduced to the George Reserve in Michigan, expressed truly exponential growth.

In 1928, 2 bucks and 4 does were introduced into the 480-ha deer-proof fenced enclosure. By 1934 there were 164 deer. During the 6-year period, growth of the deer population was unrestrained by a change in birth or death rates acting under the influence of population size. In this case, the constant, *r*, was the only factor determining the rate of change in the deer population.

Thus, for those few years, a population of increasing size grew at an increasing rate. Other rapid growth rates have occurred among ring-necked pheasants (*Phasianus colchicus*) introduced to Protection Island, Washington, and European reindeer (*Rangifer tarandus*) introduced to St. Paul Island off the Alaska coast, but these populations showed less than the maximum growth that might be expected under totally favourable circumstances.

Almost any population may approach an unrestricted rate of growth if reduced to a low level, but obviously no population can increase exponentially for very long. The supply of food may not meet the demand of the ever-increasing population; space or cover availability may be limiting; predators may respond to the large numbers of prey; or disease may spread.

Either birth rates decline, death rates increase, or both, so that eventually the population must stop growing. The greater the size of the population, the greater its dampening effect on the growth of the population. This effect has been mathematically defined and applied to the growth equation as follows:

$$\frac{\Delta N}{\Delta t} = rN\frac{(K-N)}{K} \tag{4}$$

where *K is* defined as the maximum number of individuals the environment can sustain. As the population (*N*) approaches *K*, *K* — *N* approaches zero so that when a population gets very large relative to the number the environment can sustain, its growth rate becomes nearly zero. That is, its potential growth rate (rN) is multiplied by the factor $(K - N)/K$.

For example, suppose the same population we considered earlier with an r of 0.25 has 990 individuals and the maximum number supportable by the environment is 1000:

$$\frac{\Delta N}{\Delta t} = 0.25(990)\left(\frac{1000-990}{1000}\right)$$
$$= 247.5(0.01)$$
$$= 2.5$$
$$N_{t+1} = N_t + \frac{\Delta N}{\Delta t} = 990 + 2.5 = 992.5$$

Instead of the population growing by 247 individuals as it would without any limitations, it grows only by 2.5 individuals, owing to limitations placed upon it by the finite environment. Equation (4) is known as the logistic equation. The curve it produces *is sigmoid* (S-shaped) and is illustrated in Figure elsewhere in this chapter.

Populations may sometimes exceed the maximum number that can be sustained by their habitat. In such an occurrence the term $(K - N)$ is negative. Therefore, $\Delta N/\Delta t$ is negative, resulting in a decrease in numbers.

The term K often is referred to as the *carrying capacity*. One must keep in mind that carrying capacity for animals can change from time to time as food production, cover availability, water availability, and other environmental factors vary with the seasons and successive years.

Factors such as territorial behaviour and response to crowding may interact with these external factors, so that the growth of a population may slow down before food, water, or cover shortages are measurable in the habitat. A factor that causes higher mortality or reduced birth rates as a population becomes more dense is referred to as *a density-dependent* factor.

That is, if the probability of an individual being born or surviving is lower as the numbers of animals in the population become higher, a density-dependent factor is acting to restrict population growth. Such factors include food supply, predation, disease, and territorial behaviour. There are few density-independent factors, and they are mainly related to weather, such as cold, rain, and floods.

Usually, populations in the central portions of the geographic range

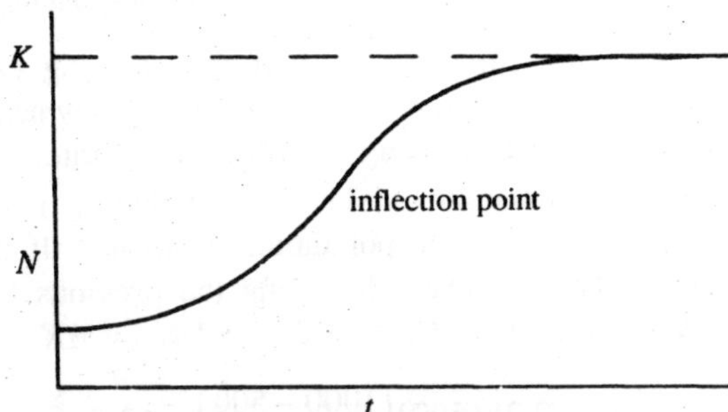

Figure 12.2: Growth of a population with a maximum number of individuals that can be sustained by the environment. $\Delta N/\Delta t = rN(K - N) K.$

of their species are limited by density-dependent agencies. Near the periphery of the range, however, where habitat may be marginal and where random weather fluctuations may exceed the tolerance of nearly all animals in the population, density-independent factors may control population numbers.

Such is the case with bobwhite in the northern parts of their range, where severe winters almost invariably result in a drastic reduction in quail numbers. Lack documented a similar effect on herons wintering in England.

The logistic equation is based purely on the operation of density-dependent factors. If a fatal flood or snowstorm strikes at some point, the irregularity in population growth will not be explained by the logistic equation.

The logistic equation, for the wildlife biologist, is useful in illustrating general principles of population growth as well as the theoretical effect of carrying capacity on reducing or stopping population expansion. The early portion of the sigmoid curve can serve as a theoretical model with which the manager can compare the growth rate of the population being managed.

A value for r can sometimes be obtained from knowledge of natality and mortality of the species under ideal captive conditions or from potential birth rates and longevity information in the field. If population growth is close to the growth rate predicted by the early phases of the logistic equation, there is little the manager can do to increase growth of the populations.

There is a further practical application of the logistic equation. If a student runs through the equation, setting N at various levels, it will be discovered that the largest value of $\Delta N/\Delta t$ is obtained when N

is half of the carrying capacity. At half the carrying capacity, an inflection point occurs in the growth curve, the point at which population growth changes from an increasing to a decreasing rate.

In managing for maximum yield of a population, it is therefore desirable to attempt to keep the population at about half the level of the carrying capacity. For example, using the previous hypothetical population with *K* of 1000 and *r* of 0.25, and *N* = ½*K:*

$$\frac{\Delta N}{\Delta t} = 0.25(500)\left(\frac{1000-500}{1000}\right) = 62.5$$

The maximum number of individuals that can be produced in a unit of time is about 62. At any other *N*, the number produced is fewer.

The factor, (*K* - *N*)/*K*, in the logistic equation, has no biological meaning or influence by itself. It must, in a real population of animals, represent some modification of birth rates or death rates. Therefore, the field biologist usually seeks an explanation for increasing or decreasing numbers of animals by examining the ratio of birth rates to death rates and the reasons for irregularities in either or both.

Most animals dealt with by wildlife managers reproduce seasonally, producing an annual spurt of offspring. Only if we stand back and view such a population over 50 or 100 years would these spurts become less visible.

In such a long period of time, comparisons with the logistic equation could be made by wildlife biologists; but management results frequently are required in a much shorter time and the wildlife manager must use other methods to assess the health of a population and evaluate its growth.

Field Studies

Returning to the African antelopes, Wilson and Hirst (1977) noted that population limitation could be effected th ough poor reproductive performance brought about by physiological factors or by increased mortality among juveniles or adults.

They set about to identify the main factors that were impeding the growth of the antelope populations on the Transvaal Nature Reserves first by reviewing available literature on their food and cover. The animals were found to be specific in their feeding and habitat requirements.

Because the soil on several of the nature reserves was poor, the researchers decided to concentrate on feeding habits, food and nutrient availability, habitat selection and availability, and seasonal body conditions, especially of breeding females.

With these factors in mind, they moved into the field, particularly the Percy Fyfe Nature Reserve, a primary study area that contained both sable and roan antelopes, and three other reserves in Transvaal. A reserve in Rhodesia (now Zimbabwe), where these two species thrived, served as a basis for comparison.

Every day for 7 months, with binoculars and a tough vehicle, they made detailed observations of two herds of antelope, one of each species. For 3 years beyond that, antelope on two reserves were watched intensively. When a calf would disappear, a team of 16-20 searchers would comb the area to find it, dead or alive.

Blood samples were taken from captured living animals; and, with recently dead animals, digestive tract contents were collected, and smears and sections were taken from the spleen, liver, kidney, lung, heart, adrenals, lymph glands, and brain. Such organs and tissues were examined for lesions and for parasites.

Healthy live animals were immobilized with a tranquilizer gun; they were then examined and fitted with coloured collars for individual identification. Species and densities of vegetation were recorded and soil fertility was assessed in the search for specific factors that caused the antelope populations to decline.

Among sable antelope studied intensively on two reserves, pregnancy rates were 100 percent when a bull was present during the May-July mating season. In a herd in which females were divided and a bull was not in full-time attendance, only 50 to 70 percent of the potentially breeding females were pregnant.

Even where pregnancy rates were high, however, antelope less than two years old suffered high mortality. Wilson and Hirst found that these young animals carried heavy infections of four different protozoan parasites. Although these parasites appeared to be important in causing the death of the young antelopes, further study revealed that parasites were fatal only if the young antelopes were in poor nutritional condition.

The researchers then compared body condition, as indicated by body weight, blood plasma proteins, packed cell volume of blood, and albumen content of the blood, between the antelopes from the Transvaal, where populations were doing poorly, and those of Rhodesia (Zimbabwe), where populations were thriving. The Transvaal animals were in very poor conditon.

This led Wilson and Hirst to examine the Transvaal range, analyzing nutrient availability in soil, water, and vegetation collected from preferred feeding areas. They also analyzed animal tissues (liver, blood, and milk). Deficiencies in phosphorus, selenium, and protein were found in food plants during the critical dry season.

On one reserve, Wilson and Hirst found, by examining digestive tract contents and observing the feeding behaviour of large herbivores, that competition from zebra (*Equus spp.*), waterbuck (*Kobus ellipsiprymnus*) f, and impala (*Aepyceros melampus*) limited the amount of food available to sable antelopes.

Roan antelope bred throughout the year, with a gestation period of about 275 days and estrus occurring 2-3 weeks after parturition (giving birth). Thus, female roan antelopes could give birth to a calf every 10-10.5 months. Sable antelope, on the other hand, had a shorter gestation period (240-248 days) but bred only once a year. Bulls were capable of producing viable sperm at 16-18 months.

In smaller reserves, the dominant herd bull of roan antelope sometimes killed young maturing bulls, thus disrupting breeding and reducing the population. The breeding success of sable antelope was satisfactory.

Wilson and Hirst found that although numbers of sable antelope on the relatively small Transvaal reserves were not high (less than 100 animals), the densities there (up to 9/km^2) were as high as, or higher than, those of good populations in Rhodesia (Zimbabwe).

Given the competition for food and the lower quality of range, the biologists concluded that a decline in antelope numbers in the Transvaal would be expected. They recommended that two Transvaal reserves be abandoned for conservation of sable antelope, on the basis of nutritional and space inadequacy.

In other reserves they recommended the use of salt licks to provide lacking minerals, and they suggested fertilizing and burning the range to enhance protein availability in forage. They also advised maintaining lower densities of antelope to reduce strain on the limited food resources.

BIRTHS AND DEATHS

The antelope study of Wilson and Hirst (1977) illustrates the elements of an excellent study of a population. To achieve its management objectives, such study involves investigation of adequacy of habitat, specifically nutrition in this case, and its effects upon birth and death rates.

A population grows according to the simple equation:

$$r = b - d$$

where

r = actual growth rate of the population

b = birth rate

d = death rate

In some populations, animals moving in or dispersing from a population may also play a role in its growth rate. The equation then becomes:

$$r = (b - d) + (i - e)$$

where

i = immigration

e = emigration

A rate represents a change per unit time. Growth rate is the number of individuals added per individual in the population per week, per month, or per year. For example, suppose 3000 young are born each year in a population of 1000 cottontails.

This represents a per-head birth rate of 3.0. During the same period, to have a stationary population, there must be a per-head death rate of 3.0, which means that for every individual present in the spring population three must die over the course of the year.

This would offset the per-head birth rate of 3.0. The population growth would then be zero ($r = 3.0 - 3.0 = 0$). Birth rates and death rates differ with age structure and sex ratios of populations.

If there are relatively many females of prime reproductive age, a population naturally will reproduce faster than one that has few females at such an age.

Therefore, the wildlife biologist, to understand population growth, should consider the following seven characteristics pertaining to birth rates:

1. Age of sexual maturity of both males and females.
2. Length of the gestation period.
3. Sex ratios.
4. Whether the species is monogamous or polygamous.
5. Number of females that breed at each age.
6. Number of young per female of various ages.
7. Influence of nutritional condition on reproduction.

Sex Ratios and Mating Systems

Sex ratios express the relative abundance of each sex in wildlife populations. The ratios usually are expressed in one of two ways. The first tells the number of males per 100 females.

The second expresses the percentage of males and females per 100 individuals in the population, with the percentage of males appearing first (e.g., in the ratio 60:40, the population consists of 60 percent males). We shall adopt the latter expression.

Sex ratios may change within populations either because of unbalanced sex ratios at birth, or, more frequently, because of sex-specific mortality associated with age. Causes may be external forces such as hunting or the vulnerability of incubating females to predators.

Johnson and Sargeant (1977) presented strong evidence that red foxes (*Vulpes vulpes*) killed enough hens during the nesting season to distort the sex ratio in a large population of mallards (*Anas platyrhynchos*).

Some inherent mechanisms apparently cause differential mortality during embryonic development in some species, under conditions that are not always clear. Bellrose et al. (1961) recognized these age-related differences and adopted the following categories:

Primary Sex Ratio

The sex ratio at fertilization; normally 50:50 based on simple statistical probability.

Secondary Sex Ratio

The sex ratio at birth or hatching; usually approximates 50:50, but may show the first indication of sex-specific mortality (e.g., 49:51). At birth, the fawns of white-tailed deer suffering from nutritional stress may exhibit a ratio favouring males by as much as 72:28.

Tertiary Sex Ratio

The sex ratio of juveniles; important because it indicates the proportion of each sex later entering the breeding population; for game species, hunting becomes an external influence for the first time; some inherent species-specific differences also may be present.

Quaternary Sex Ratio

The sex ratio of the adult population; often clearly skewed in favour of one sex; some species of diving ducks (e.g., redheads, *Aythya americana*) are heavily unbalanced in favour of males; in most populations of large ungulates, such as deer or bighorn sheep (*Ovis canadensis*), females predominate because hunting pressure normally selects males with antlers or horns.

Waterfowl and other wildlife follow one of three types of mating systems. These systems interact with population sex ratios in ways that profoundly affect annual production. Mating systems include the following:

Monogamy

a. Seasonal. Pair-bonds established only for the current breeding season (e.g., pintails, *Anas acuta*, and other dabbling ducks)

b. Life-time. Pair-bonds established for as long as both mates remain alive (e.g., coyotes, *Canis latrans*, and several other canids; Canada geese, *Branta canadensis*, and allied species)

Polygamy

a. Polyandry. Several males per female; extremely rare in most groups of vertebrates but occurs in a few birds (e.g., Wilson's phalarope, *Steganopus tricolour*)

b. Polygyny. Several females per male (e.g., ring-necked pheasants; elk, *Cervus elaphus canadensis*)

Promiscuity

Indiscriminate mating (e.g., cottontails, *Sylvilagus floridanus;* bobcats, *Felis rufus* and many other felids)

The effect of quaternary sex ratios on production in monogamous species is shown in the following example; the maximum number of nests serves as an arbitrary measure of production:

Sex Ratio	*Maximum Number of Nests per 100 Birds*
50:50	50 (100% production)
60:40	40 (80% production)
40:60	40 (80% production)

The example illustrates the fact that monogamous species require a balanced sex ratio for the maximum production of offspring. Any deviation favouring either sex reduces prpduction. Thus, in normal situations, hunting regulations for such species probably should not set sex-specific bag limits.

However, when a sex ratio is highly imbalanced, hunting regulations can focus the harvest on the more abundant sex. For example, the harvest of ducks in many states is governed by a point system. That is, the bag limit is based on the accumulation of100 points instead of on a set number of birds.

Because the proportion of females is low in some species or populations, the point system places high values on hens and low values on drakes as a means of shifting the shooting pressure from one sex to the other.

In polygynous species, the situation is quite different. Females represent a premium for increased production. Males, within reason, become expendable, as follows:

Sex Ratio	*Maximum Number of Nests per 100 Birds*
40:60	60 (100% production)
50:50	50 (83% production)
60:40	40 (66% production)

Thus, in polygynous species, 100 percent production occurs at any ratio where females make up more than 50 percent of the population (the 40:60 ratio was selected arbitrarily for comparison with production at the same sex ratio for a monogamous species, shown earlier). Indeed, the disparity favouring females might be increased to 30:70 or more for even greater production of offspring per 100 adults.

Eventually, however, the point is reached where males no longer can successfully court and mate with such a large number of females. At such a point, the imbalance becomes so great that the "extra" females no longer contribute offspring and thus become an expendable surplus.

Among penned ring-necked pheasants, McAtee found that some hens remained unmated when the sex ratio reached 12:88. In most circumstances, however, male pheasants can be heavily harvested each autumn without endangering production the following spring.

Age-Specific Birth Rates

The number of births per individual (or, more commonly, per 1000 individuals) in a population is called the *crude birth rate*. The crude birth rate reveals no details about what age groups actually contribute offspring to the population.

Birth rates are by no means fixed within species or within populations, and variations in natality may account for large shifts in the sizes and densities of wildlife populations.

The number of offspring produced during a particular period depends upon the number of females in each age class, the number of these that actually mate, and the fecundity of each age class (the same considerations may apply to males, of course, but females generally govern age-specific production). The number of offspring per female expressed by age classes is called the *age-specific birth rate*.

As an example, a population of brook trout (*Salvelinus fontinalis*) breeding in Lawrence Creek, Wisconsin, showed large age-specific differences in egg production. Reproduction relied heavily on females in age classes I and II (1.5- and 2.5-year-old trout), which together produced more than 98 percent of the eggs.

Three facts emerge from the details of the study. First, egg production per mature female increased with age. Yearling trout produced an average of about 425 eggs per mature female, whereas 2- and 3-year-old females each produced an average of 617 and 906 eggs, respectively.

Second, because not all of the females in age class I were sexually mature, the average egg production for all females (mature and immature combined) in age class I was reduced even further (to 354 eggs). Third, even with lower egg production per female, the individuals in age class I were so numerous that the group still accounted for more than 75 percent of all egg production.

These facts have management implications. In this case, Lawrence Creek is heavily fished, and few trout survive beyond age class III even though the population remains relatively stable. Thus, armed with knowledge of the age-specific reproductive rate, managers can adjust the fishing regulations in ways that will help protect the breeding population (e.g., changes in creel limits, season lengths, size limits, or a combination of these). With reduced fishing pressure, more females would reach the older age classes and thereby would contribute more to egg production.

Table 12.1: Number of Eggs Produced by Brook Trout in Lawrence Creek, Wisconsin, 1955-56.

Age Class (Year)	*Percent Sexually Mature*	*Total Number of Eggs*	*Percent of Eggs Contributed*
I	83.3	1,706,685	76.6
II	100.0	485,756	21.8
III	100.0	32,014	1.4
IV	100.0	2,318	0.1
V and VI	100.0	2,555	0.1
Total		**2,229,328**	**100.0**

Such a shift would result in greater production from the trout population in Lawrence Creek, because older females are all sexually mature and individually can produce larger numbers of eggs than younger females.

In another example, Woolf and Harder (1979) compared age-specific birth rates among white-tailed deer in three states. These data revealed poor reproductive success in the Pennsylvania herd, which was densely populated and maintained with artificial food, in comparison with herds in Iowa and Ohio.

Several hypotheses were explored to explain the poor reproductive performance in the Pennsylvania deer. The factors that were investigated included adrenal stress from heightened social interactions in the populous herd and reduced ovulation rates induced by chemicals in the diet of acorns. Ultimately, however, the low reproductive rates in the Pennsylvania herd were attributed to poor summer nutrition.

This factor extended the lactation period into the autumn breeding season. Thus, fawns were weaned unusually late in the year and were unprepared to breed in the autumn. Moreover, because the older does were still lactating, their nutritional reserves were so low that ovulation and early fetal development were impaired.

These conditions lowered the ovulation rates and induced early mortality in the fetuses of older does. Woolf and Harder (1979) estimated ovulation rates using the common method of examining *corpora lutea.* These structures are the remains of ovarian follicles from which eggs are produced by sexually mature females during the current breeding season. To determine the number of corpora lutea, wildlife managers examine thin cross sections of the ovaries taken from females collected on the research area.

The corpora lutea appear as small, yellowish spheres embedded in the ovarian tissues. Each corpus luteum represents the ovulation of a single egg, and the ratio of young fetuses to corpora lutea thus indicates the fertilization rate. Using these and other techniques, wildlife managers can obtain accurate measures of the reproductive contributions of each age class in a population.

A table of age-specific birth rates thus becomes useful for depicting the reproductive performance of a population. A few examples are shown in Table elsewhere in this chapter.

As an illustration of how reproduction may vary with age structure, consider the population of Canada geese listed in Table elsewhere in this chapter, and two different distributions of breeding birds within the age classes. In the first case, the age structure is represented by 34 percent 1-year olds, 33 percent 2-3-year olds, and 33 percent 4+-year-old birds. Egg production from 100 females would be (34 × 0) + (33 × .20 × 4.6) + (33 × .72 × 6.4) = 182 eggs.

If the age ratio were to change to 40:40:20, respectively, for each of the age classes, then the number of eggs laid by 100 females would be (40 × 0) + (40 × .20 × 4.6) + (20 × .72 × 6.4) = 100 eggs. Thus, with a change in age structure as described—which might easily occur in just a few years-egg production can be diminished significantly [here, by (182 - 100)/182 = 45 percent].

Additive and Compensatory Mortality

Animal mortality the losses from a population-may be considered as either *additive* or *compensatory*. Numerous environmental factors, including disease, malnutrition, predation, and severe weather, act on members of a population.

Table 12.2: Comparison of Reproductive Performance in Three Populations of White-Tailed Deer.

Area	*Percent Fawns Pregnant*	*Percent Adult Does Pregnant*	*Corpora Lutea per Pregnant Doe*	*Fetuses per Pregnant Doe*
Rachelwood Wildlife Preserve, Pennsylvania	0.0	93.5	1.60	1.40
Plum Brook Station, Ohio	0.0	94.8	1.95	1.77
DeSoto National Wildlife Refuge, Iowa	83.6	100.0	2.23	2.10

For example, given a population of 100 animals, food shortages and disease acting together might have the potential of removing 40 individuals during the course of several months. At the same time, however, predators also might have the potential of removing 40 animals.

If these factors-starvation, disease, and predation—were *additive*, then a total of 80 animals would die from the combined action of these forces. But it is unlikely that the mortality would reach such a level.

Competition for food is reduced whenever predators remove some animals from the population. As starvation lessens, so too does the incidence of disease-related mortality, and fewer animals actually die

Table 12.3: Some Examples of Age-Specific Laying and Birth Rates.

Species and Source						
Brook trout	Age (years)	0.5	1.5	2.5	3.5	4.5
(*Salvelinus fontinalis*)	Percent breeding	0	83.3	100	100	100
McFadden (1961)	Number of eggs per breeding female	0	425	617	906	1196
Canada goose	Age (year)	1	2-3	4+		
(*Branta canadensis*)	Percent breeding	0	20	72		
Cooper (1978)	Number of eggs per breeding female	0	4.6	6.4		
Blue whale	Age (years)	0-3	4-5	6-7	8-11	12+
(*Balaenoptera musculus*)	Number of calves per breeding female	0	0.19	0.44	0.50	0.45
Usher (1972)						
White-tailed deer	Age (years)	0.5	1.5	2.5	3.5	4.5
(*Odocoileus virginianus*)	Percent breeding	16	68	77	81	84
Teer et al. (1965)	Number of fawns per breeding female	0.88	1.32	1.52	1.52	1.52

from malnutrition and sickness because of the interaction with predation. If predators remove 30 animals, only 10 might die of disease. Thus, the mortality factors act in *a compensatory* way.

Errington concluded that mink (*Mustela vison*) preyed on muskrats that were "surplus" members of a crowded population. The surplus muskrats were "social outcasts," because they could not obtain and hold breeding territories.

As such, the outcasts were vulnerable to diseases and predation. If mink did not kill them, the surplus muskrats soon succumbed to disease. As Errington stated, "victims of one agency simply miss becoming victims of another" and many types of mortality are "at least partly intercompensatory in net population effect." Other histories of animal populations involving the compensatory nature of predation are described in Of *Predation and Life*.

Density-dependent factors operate in a compensatory manner. By contrast, severe weather may act in an *additive* way. That is, even though a population may be reduced to low levels, a sleet or ice storm may kill a fixed proportion of the original number of animals, regardless of how many had been taken earlier by predation or disease.

If, however, the storm preceded a period of food shortage, the effects of the bad weather might be compensatory, for the remaining animals likely would experience increased survival.

Hunting is a common form of mortality in populations of game species. Hunting mortality is frequently compensatory, because it usually increases the life expectancy of individuals surviving the hunt, promotes higher reproductive rates, or does both. Swenson compared the reproduction of mountain goats (*Oreamnos americanus*) subjected to different levels of hunting mortality.

One result of the study indicated that the summer ratio of kids to older goats (yearlings and adults) was higher after hunting had reduced the number of older goats the previous autumn. Decreased competition for winter forage explained the increased production of kids. Some caution was expressed, however, because the goats in this study were members of a population that had been introduced 10-20 years before, and the population was still expanding in logistic growth.

Nevertheless, logic suggests that reduced competition in any population will increase the survival of the remaining individuals and will enhance birth rates.

Wagner et al. summarized the compensatory nature of hunting in mathematical terms in this way: The addition of a given pecentage of mortality (from such factors as hunting or fishing) does 1ot add one-for-one with the existing annual mortality from other causes (e.g., disease). Instead, the total annual mortality increases by a much smaller percentage than is measured by the actual pecentage of individuals removed by hunting alone.

The formula for calculating the crude annual mortality rate is

$$a = m + n - mn$$

where

a = crude annual mortality rate
m = mortality rate from hunting or fishing
n = natural mortality rate

In a population with a natural annual mortality rate of 70 percent, the addition of 20 percent mortality from hunting would not increase the total mortality to 90 percent, but only from 70 to 76 percent:

$$a = 70\% + 20\% - (70\% \times 20\%) = 76\%$$

In other words, hunting removes some animals that otherwise would die from natural causes. Errington's ideas about surpluses do not apply here. Instead, the relationship shown in the example emerges simply because an animal can only die from one of the types of causes to which it is exposed: natural mortality or hunting mortality.

The natural mortality rate remains the same with or without hunting, although the actual number of animals dying from natural causes is less where hunting first removes a fraction of the population. Moreover, a given percentage of harvest increases a small annual mortality rate more than a !arge one.

For example, a 20 percent harvest raises a 40 percent annual mortality rate to 52 percent—a 30 percent increase-whereas, as we have shown, the same 20 percent harvest raises a 70 percent annual mortality rate to 76 percent, for only a 9 percent increase.

Such a relationship, in part, accounts for the more visible effects of hunting species with low mortality such as big game and geese. The relationship supports Hickey's conclusion that the ability to withstand harvest is a function of each species' annual mortality rate.

Life Tables and Survivorship Curves

Comparisons of mortality between populations can be made by use of *life tables* and *survivorship curves. A* life table is a systematic means of describing mortality as it affects various age groups in a population. Deevey published a classic review of life tables for natural populations of animals. Murie described mortality of Dall sheep (*Ovis dalli*) in Mount McKinley National Park, Alaska. Over a period of several years he collected skulls of sheep found in the park.

The horns of sheep grow in annual bursts, leaving a ring between each annual increment so that the age of a Dall sheep can be estimated by counting annual growth segments. From 608 sheep Murie constructed the life table shown in Table elsewhere in this chapter.

The columns of a life table are as follows:

x = An appropriate time interval.

l_x = The number of animals living at the beginning of interval x. It is traditional to convert whatever sample size one has to 1000 at the beginning of the l_x column, representing 1000 animals born or hatched.

d_d = The number of animals dying during interval x.

1000 q_x = The proportion of animals that die per interval x. It is computed as follows: 1000 qx

= $(d_x \div l_x) \times 1000$.

e_x = The life expectancy expressed as the number of additional intervals an individual animal can expect to live at the beginning of interval x.

The derivation e_x involves a few extra calculations, as follows:

$$e_x = \frac{T_x}{l_x}$$

and

$$L_x = \frac{l_x + l_{x+1}}{2}$$

or the average number of animals alive at the midpoint of an interval x.

T_x is the sum of the L_xs from the bottom of the table up through the desired x interval.

Biologists often are asked questions such as, "How long does a robin live?" or "How long does a mule deer live?" The answer to such questions depends upon what age the animal has achieved.

The American robin (*Turdus migratorius*) can live to be 7 years old, but the probability of a newly hatched robin doing so is much less than 1 percent. Early mortality, in fact, is so high among songbirds that most life tables for them do not begin until late in the summer or early fall of a bird's first year.

Once a robin has lived to November 1, on the average it will live another 1.37 years. Over half of them will die in the next year. Murie's life table for Dall sheep shows that life expectancy is 7.1 years for a newborn lamb. There was a flaw in Murie's data, however, in that the skulls of very young lambs that died possibly were consumed totally by scavengers or predators.

Thus, young animals would be underrepresented in his life table and life expectancy at birth would be overestimated. Once a sheep reached the age of 6 years one may see that it can expect to live, on the average, another 3.4 years, to reach the age of 9.4 years.

A survivorship curve is constructed by plotting the l_x column of a life table against time. A classic use of survivorship curves in wildlife management was described by Taber and Dasmann for black-tailed deer (*Odocoileus hemionus columbianus*) *in* two habitats in California.

By use of survivorship curves, the researchers were able to compare survival rates of both sexes of five populations of ungulates. All survivorship curves were found to decline steeply in the first year, indicating high death rates of young animals.

After that, considerable variation among populations occurred. (The logarithmic scale is used on the abscissa to expand the lower parts of the curve. On a logarithmic scale, the removal of a constant proportion of animals would result in a straight declining line, such as occurs in female red deer (*Cervus elaphus*) from age 3 through 14.

Survival of male black-tailed deer was lower than for females because o(selective hunting of males. Taber and Dasmann noted from their survivorship curves that survival of black-tailed deer was much less for both sexes in shrubland than in chaparral. By examining other population features, they found that fawn production in the shrubland deer was higher than in the chaparral deer (0.76 fawns per adult doe in shrubland to 0.53 per doe in chaparral in December).

This higher production in the shrubland resulted in more intense competition for food, poorer nutrition, and higher mortality. Chaparral is a mixture of woody shrubs, while "shrubland" consists of scattered shrubs and herbs. In the shrubland, deer food was more abundant, and this feature permitted deer to maintain higher densities (25 per km^2) than in chaparral (11 per km^2).

Survival rates in the shrubland, however, were lower than in chaparral, presumably because the shrubland was fully stocked and the higher birth rates resulted in consequent higher losses through starvation. Once the carrying capacity of the shrublands had been attained, a proportion of deer that were not removed by hunting starved, so that the

Table 12.4: Life Table for the Dall Mountain Sheep (Ovis dalli) Based on the Known Age at Death of 608 Sheep Dying Before 1937 (Both Sexes Combined).

x	x'	d_x	l_x	$1000\ q_x$	e_x
Age (years)	*Age as Percent Deviation from Mean Length of Life*	*Number Dying in Age Interval Out of 1000 Born*	*Number Surviving at Beginning of Age Interval Out of 1000 Born*	*Mortality Rate per Thousand Alive at Beginning of Age Interval*	*Expectation of Life: or Mean Lifetime Remaining to Those Attaining Age Interval (years)*
0-0.5	–100.0	54	1000	54.0	7.1
0.5-1	–93.0	145	946	153.0	—
1-2	–85.9	12	801	15.0	7.7
2-3	–71.8	13	789	16.5	6.8
3-4	–57.7	12	776	15.5	5.9
4-5	–43.5	30	764	39.3	5.0
5-6	–29.5	46	734	62.6	4.2
6-7	–15.4	48	688	69.9	3.4
7-8	–1.1	69	640	108.0	2.6

(Table 12.4 Contd.)

(Table 12.4 Contd.)

x	x'	d_x	l_x	$1000\ q_x$	e_x
Age (years)	***Age as Percent Deviation from Mean Length of Life***	***Number Dying in Age Interval Out of 1000 Born***	***Number Surviving at Beginning of Age Interval Out of 1000 Born***	***Mortality Rate per Thousand Alive at Beginning of Age Interval***	***Expectation of Life: or Mean Lifetime Remaining to Those Attaining Age Interval***
8-9	+13.0	132	571	231.0	1.9
9-10	+27.0	187	439	426.0	1.3
10-11	+41.0	156	252	619.0	0.9
11-12	+55.0	90	96	937.0	0.6
12-13	+69.0	3	6	500.0	1.2
13-14	+84.0	3	3	1000.0	0.5

growth rate in both shrublands and chaparral was essentially zero.

Taber and Dasmann also pointed out that dynamics of different populations of the same species may vary widely from place to place. It is therefore difficult to say that birth and death rates of a particular population are "typical" of a species; differences among populations of the same species reflect different environmental conditions.

There also is evidence that the genetic makeup of a population may change with the passage of time or in response to some environmental factor. (Such change is the basis for the theory of evolution.) Some changes can take place quite rapidly.

For example, European rabbits (*Oryctolagus cuniculus*) in Australia were in the 1950s intentionally infected with myxomatosis, a viral disease, to control their numbers. The rabbits declined rapidly. There are still rabbits in Australia-not as many, but they are genetically more resistant to the virus than the rabbits in Australia in the 1940s.

Other less noticeable and unmeasured genetic changes possibly occur in many populations, changes that influence their birth rates and death rates, and that may be responsible for population irruptions and crashes for which no external cause may be apparent.

Sources of Population Data

Obtaining accurate information about animal numbers and densities remains one of the more difficult and challenging tasks for wildlife managers. Data are gathered in the form of *censuses*, *estimates*, and *indices*. *A* census is a complete count of individuals in a population.

Some species, particularly those living in open areas, may be counted from aircraft (e.g., pronghorns, *Antilocapra americana*, or winter flocks of waterfowl). Complete counts also are possible in a few special situations. Whooping cranes (*Grus americana*) are large, white birds, and thus are highly visible against a backdrop of marsh vegetation; the small population can be counted accurately in either summer or winter.

A complete count is rarely possible in areas where vegetation or topography conceals animals, or where the population is quite large. In such cases, an estimate may be made on the basis of a statistical sample. A sample may be obtained by counting inanimate objects (e.g., droppings, nests, and dens) or by counting animals.

In either case, the sample is taken on a plot or transect of a known size. Other information often is required before the sample can be interpreted. In the case of dens or burrows, for example, the average family size must be determined, then multiplied by the number of active burrows in the plots. Population estimates for prairie dogs (Cynomys spp.) are made in that manner, as are estimates based on the lodges of beaver (*Castor canadensis*). For droppings, the average number of droppings per day per animal must be known, as well as the maximum number of days since the droppings were deposited.

A common way of sampling living animals is based on a capture-recapture ratio. The estimate requires capturing, marking, and releasing a known number of animals, then resampling the population later. The population size is estimated using the proportion of the marked animals either recaptured (or resighted) in the second sample:

$$\text{No. in population} = \frac{\text{no. marked \& released} \times \text{no. resampled}}{\text{no. marked in resample}}$$

The capture-recapture method, known as the Lincoln or Petersen Index (although the ratio actually is an estimate), often underestimates the population because of a higher likelihood of recapturing (or sighting) marked animals than unmarked animals.

An index is a quantitative measure of a population. However, it seldom provides exact numbers or even an estimate of the numbers or densities of animals in a population. Indices instead compare relative abundance between areas, or changes in abundance from one time to another in the same area.

Counts of displaying male woodcocks (*Scolopax minor*) and ruffed grouse (*Bonasa umbellus*) along established routes are commonly used indices. Others include counts of pheasants (*Phasianus colcbicus*) and red foxes (*Vulpes vulpes*) by rural mail-carriers.

Biologists also obtain sex and age data from various sources. The age structure of fish populations often is determined from a collection of scales; these show annual growth rings known as *annuli*. In autumn, hunters voluntarily submit wings or tails from various kinds of game birds for analysis; quail, grouse, and other birds are sampled for year-to-year changes in the ratio of juveniles to adults.

Waterfowl are among the migratory birds sampled by the U.S. Fish and Wildlife Service, as thousands of duck wings are analyzed each year for age and sex ratios, by species, in what are known as *wing bees*. At roadside check stations, state biologists obtain age and sex information from big game killed by hunters.

Tooth replacement and wear are among the features commonly inspected for age determination in deer and other large mammals. However, the methods for determining age and sex vary greatly by species. Methods range from studying the shapes of feathers among birds to determining the weight of eye lenses for some kinds of small mammals; these methods are reviewed in detail by Larson and Taber.

Sampling theory must be considered when population data are collected. The subject is complex, and an extensive body of literature may be consulted. In its simplest form, however, sampling theory requires that the collected data represent the population at large.

For example, the age distribution shown by scales collected from 100 fish in a lake produces a statistical estimate of the entire fish population in the lake. What is sampled (e.g., 100 fish) in theory represents what was not sampled (e.g., all of the other fish in the population).

However, one or more sources of *sampling bias* may affect the reliability of the samples, and hence the results of the analysis may not be accurate. If the fish population was sampled with a gill net, then the smaller—and thus the younger—fish likely escaped capture. Hence, scales would be collected only from the larger—and older—fish, and the analysis would reflect a population lacking young age classes.

In this case, the bias results from the equipment and methods used in the field. Similarly, both age and sex data collected from deer at roadside check stations are usually biased (i.e., few does and fawns are shot), so those data usually reflect only the adult male segment of the adult population. In this case, the sampling bias is associated with the selectivity of hunters for certain age and sex groups.

However, most population data can be corrected or applied in ways that compensate for sampling bias; alternate field methods or equipment also may reduce the bias to acceptable levels (e.g., electrofishing equipment, or "fish shockers," may yield a better sample of size and age classes than gill nets provide). The important matter here, however, is that bias must be recognized and addressed before population data can produce sound management.

Organization of a Population Management Problem

A wildlife manager charged with solving a problem of population ecology may be faced with a somewhat bewildering array of possible causes of an "unsatisfactory" performance by the animals he or she wishes to manage. The following outline is intended to suggest a means of organizing efforts. The objective, of course, is to identify those factors that are most responsible for preventing the further growth of the population. These factors impede births, increase deaths, or both.

A. Extrinsic Factors

1. Density-independent (primarily weather conditions)
   a. Cause of direct mortality?
   b. Center or periphery of species range?
   c. Does weather have a substantial influence on food quality or quantity-which are density dependent factors?
2. Density dependent
   a. Food
      (1) Quality. Are necessary nutrients present?
      (2) Quantity. Is enough food available?
   b. Cover
      (1) Shelter from elements. Are quality and quantity sufficient?
      (2) Escape or hiding cover-for predators or from predators. Are quality and quantity sufficient?
   c. Refugia available. Are there patches of habitat in the range of the population in which animals have a high likelihood of escaping various mortality factors, such as predators, hunters, parasites, and disease?
   d. Competitors. Is there competition for resources by other species?
   e. Diseases and parasites. Are these factors influencing birth and death rates?
   f. Predators. Are predators controlling the population?
   g. Buffer species. Are other prey species present that absorb some of the impact of predation, particularly when the species being considered is at low densities?
   h. Hunting harvest. Is harvest toll replaced by the next season's production of huntable animals?
   i. Interactions among various factors. What interactions occur? Food supply-disease? Food supply-predation? Food supply-competition? Cover-predation? Buffer species-predation?

B. Intrinsic Factors

1. Genetically stable factors
   a. Litter or brood sizes. What is the inherent potential of

the species to reproduce?

b. Longevity. How long can individuals live?

c. Habitat selection for breeding, feeding, resting. Is it available according to inherent needs of the species?

d. Self-limiting factors. Does the species possess self-limiting behaviour such as territorial spacing, or restricted breeding among selected members of a group?

e. Dispersal. Is there an opportunity for immigration and emigration?

f. Interactions. How do inherent features interact, such as territorial behaviour-food supply, dispersal-food supply, birth rates-food supply?

2. Genetically variable factors

a. Birth rates. Within the physiological limits of the species, does the population show varying birth rates?

b. Survival rates. Does a population differ genetically from time to time in the ability of individuals to withstand stress, or has there been a response to a strong selective factor such as disease or biocides?

As can be seen, answering questions about a population may be a complex task. The most fruitful and simplest approach is usually to examine the more obvious factors first, such as weather, food, cover, and the behavioural nature of the animals. Should such an approach fail to provide satisfactory answers, the more subtle interactions must be examined.

Usually, the talents of a team of researchers, with various forms of expertise, must be called upon. Once the most important factors (key factors) limiting a population are identified, the manager may attempt to modify those factors.

Sometimes nothing can be done, such as when the weather or climate of a particular area is simply unsuitable for the welfare of the species; but most often the application of suitable controls on key factors will result in a desired change in the population being managed.

POPULATION MODELS

When wildlife managers predict how many mallard ducks will be present in the fall population, based upon sample counts of breeding ducks and the number of prairie ponds in the spring, they are using a model. Similarly, an estimate of the number of deer present obtained by sampling and counting pellet groups (droppings) left by the deer is another example of the use of a model. A model, as defined by Walters is any physical or abstract representation of the structure and function of a real system.

In the past decade, attempts have been made to construct models of everything biological, from cellular physiology to entire biomes. These representations are simulation models that trace through a period of time the changes that take place in a system, given some beginning conditions and some circumstances that effect changes in those conditions.

Populations of animals lend themselves reasonably well to the methods of simulation modeling. Owing to the speed with which they perform calculations and the convenience they offer for altering variables at various stages in the operation of a model, computers offer wildlife biologists new opportunities to simulate populations of animals under study and to predict the effects of various management procedures.

The basic tools for doing so are an understanding of FORTRAN or Pascal computer language and several years of data for a population, including age structure, sex ratios, birth rates, death rates, immigration, emigration, and environmental conditions that influence these factors.

Computer models have been developed to simulate wolf (*Canis* lupus)-moose (*Alces alces*) populations on Isle Royale and the projected recovery of whooping cranes (*Grus americana*). An excellent example of a model was constructed to simulate a population of mule deer in Colorado. These authors noted that a model has three basic values:

(1) it forces the researcher to think about population dynamics in new ways (conceptual value);

(2) the researcher must become aware of the usefulness of various types of information necessary to construct an accurate model, and therefore of the information necessary to understand population functions (developmental value); and

(3) the model may be useful in predicting future courses of the modeled population or the effects of manipulation of the population by adjusting rates of exploitation or by altering the environment (output value).

The model of the mule deer population studied by Medin and Anderson utilized the following variables gathered over several years of study: vegetation available as food, weather, food consumption, nutrients in food, other animals as predators or competitors, age, sex, and number of deer present, natural mortality rates, hunter harvest, age structure in the hunter harvest of deer, birth rates as determined by ovarian and fetal analysis, and the condition of the deer.

Only one environmental variable, the amount of precipitation in the April-July period, seemed to have a significant effect on birth rates; it did so by affecting the amount of nitrogen available in winter forage. The density of deer in winter had an inverse influence on their birth rates.

By calculating functions for these variables and feeding them into a computer, Medin and Anderson were able to simulate closely the dynamics of the deer population over 5 years for which reliable data were available. The assumption is then that the model might be projected into the future, given a reasonable assortment of May-July precipitation values.

The authors also tested the effects of different harvest strategies on the model. Then the authors asked some "what if" questions of their model and obtained results depicted in Figure elsewhere in this chapter.

The student can readily see the value of a population model, provided that the model is reasonably realistic. Biologists who develop models also can recognize the frequent inadequacies of field data and assumptions of cause and effect of variables used in the model.

A population model should be only one of many forms of information the wildlife manager may use. At this point in the development of population models, an attitude of informed skepticism seems appropriate.

THE HUMAN POPULATION

It is not possible to consider management of wildlife populations without also considering management of the human population. In 1988, there were about 5.1 billion people on the earth, and this number has been increasing at a rate of 1.7 percent per year.

At such a rate (which has prevailed over the past several decades), a doubling of human numbers will occur in 40 years. Each day, more than 200,000 people are added to those already on the earth. Only 6 of the 170 nations of the world show zero or negative growth rates.

The 1988 U.S. population of 246 million has been growing at a rate of 0.7 percent. Hutchinson (1978) observed that a glance at a newspaper suggests that many people behave as if there were no carrying capacity, no upper limit, to the number of people that the earth can support.

It is true that over the centuries human inventiveness has periodically increased the carrying capacity of the earth, first by changing the way of living from a hunting and food gathering mode of existence to one of agriculture. Later, industrialization, mechanization, and rapid transportation permitted people to trade useful items for food, and fewer farmers were needed to produce human food. The "green revolution," which resulted in the development of high yield grains, further permitted the feeding of even more people.

But agricultural production over the past decade has scarcely been keeping pace with population growth, and an estimated 20-40 percent of the people in the world are underfed.

Most ecologists believe that the carrying capacity of the planet for humans is rapidly being approached, and that resources cannot be produced and processed fast enough to meet the demands of an ever-growing population.

In some places, such as India and Bangladesh, the carrying capacity has already been surpassed. The need to feed, clothe, and house the growing billions of people requires both more extensive and more intensive use of the land and waters. In doing so, humans compete with other animals.

Faced with the necessity of raising crops and trees and mining energy sources of all kinds to supply starving people, many people find that arguments for providing habitat for wildlife lose their strength; and the wildlife management goal of balancing the needs of other animals and those of humans is weighted by the sheer numbers and "humane" priority of our own species.

The influence of the human population is truly global. Lead from the exhausts of millions of automobiles is found in the Antarctic ice pack, and chlorinated hydrocarbon pesticides are infused from the atmosphere into food chains of the tundra, where pesticides have never been applied.

Sulfur and nitrogen oxides emitted from power plants in one country cause acidification of lakes and soils in another. The production and use of resources to support humanity affect the resources for other animals, usually to the animals' detriment. There is no doubt that the human population will stop growing at some point.

The question is only whether human numbers will be controlled by such natural checks as starvation and territorial defense, or by the intelligent application of methods to reduce the birth rate to match death rates, the latter of which have been lowered through medical treatment. The future of wildlife as well as that of humans rests upon such a choice.

INDEX

D

E

F

N

O

P

R